THE DIGITAL ORGANIZATION

AlliedSignal's Success with Business Technology

JAMES D. BEST

JOHN WILEY & SONS, INC.

New York • Chichester • Weinheim • Brisbane • Singapore • Toronto

This text is printed on acid-free paper.

Library of Congress Cataloging-in-Publication Data:

Best, James D., 1945–
    The digital organization : AlliedSignal's success with business
technology / James D. Best.
        p.   cm.
    Includes index.
    ISBN 0-471-16159-4 (cloth : alk. paper)
    1. Allied-Signal Inc.—Management. 2. High technology industries—
United States—Management—Case studies. 3. International business
enterprises—United States—Management—Case studies. 4. Management
information systems—Case studies. I. Title.
HD9503.B47   1997
658'.05—dc20   96-34960

To the staff of the AlliedSignal Computing Technology Center.
Thank you for your skill, dedication, and professional attitude.

CONTENTS

CONTENTS

BEGINNINGS

Making money is art and good business is the best art.

—ANDY WARHOL

Larry Bossidy was concerned. A year earlier, a project had been approved to build a new financial information system. He intended to make this state-of-the-art system a key element in his strategy to operate AlliedSignal as "one company" on a global basis. The company needed the new computerized process to reduce the closing cycle, analyze financial trends, measure performance, and allocate corporate resources. Now he discovered the project had stalled due to technical difficulties.

As AlliedSignal's Vice President of computing and network operations, I knew Mr. Bossidy's "concern" would kick off a frenzied effort to get the project moving again. We eventually succeeded in saving most of the foundation work, but the near-disaster taught AlliedSignal that it needed to change the way it managed new computer technology. An investigation revealed far too many mistakes. Management had put too much faith in a new technical solution, did not insist on common codes for key data, selected embryonic development tools from boutique software houses, and did not install the prerequisite infrastructure for enterprise computer systems. AlliedSignal, however, had the good fortune of experiencing this problem several years ago, allowing

1

the company to learn early that network computing requires discipline and good management.

Speed Demon

Typical of Bossidy, difficulties with one systems project did not dissuade him from his determination to upgrade information systems at AlliedSignal. Instead, he figured out what went wrong and reset the company's course.

Larry Bossidy set a rapid pace for AlliedSignal that forced people to break the old rules and invent new ways of getting things done. In World War II, industry designed and produced unbelievable volumes of goods and services. Even Bossidy can't maintain the urgency of war, but he does drive AlliedSignal faster than anyone else believes possible.

Why the rush? In 1991, when Bossidy left General Electric to be CEO of AlliedSignal, he inherited a menagerie of businesses with moribund performance. Bossidy found AlliedSignal trailing behind world-class companies that were continuing to increase their competitive lead. In a merciless global economy that feasts on the weak, Bossidy needed to catch up in a hurry to survive, then continue sprinting to remain a world player. Speed alone does not make you competitive; you also need to move in the right direction. Purposeless flailing-about creates a black hole into which you pour time, money, and your best people. Bossidy's success comes from knowing where he wants to go and keeping the entire company focused on the business strategy.

Through his leadership, AlliedSignal made a quick transition from a stodgy Fortune 50 company into a premier competitor with a reputation for success. AlliedSignal's earnings have grown at a compound average of 25 percent, the share price has appreciated 279 percent, and the market valuation of AlliedSignal increased from $3.6 billion to $13.4 billion. The company earned these results by having a good business strategy, aggressively innovating new ways to achieve stretch targets, making smart acquisitions and divestitures, careful allocation of assets, meeting customer needs, and aggressively managing every aspect of the company's operations. Information technology plays a crucial role in each of these business objectives.

Bossidy insists that information technology investments leverage the organization's ability to increase sales, improve productivity, share common solutions, and operate as one company. This tightly coupled approach sets AlliedSignal apart from many other companies. A business perspective, instead of an unfocused zeal for technology, forces rational investment decisions, deployment of technical solutions to the workplace, and the engagement of general management.

AlliedSignal knows how to make information technology contribute to profits. *The Digital Organization* explains how AlliedSignal uses technology initiatives to meet corporate goals. As AlliedSignal's Vice President of Computing and Network Operations, I was responsible for these initiatives. This book presents the perspective of practitioners held accountable for delivering information technology to a company driven to exemplary performance.

The Digital Organization

There is nothing mysterious nor magical about computers. They can be managed as easily as other aspects of the corporation—if you pay attention and manage consistently. Bossidy directed his business leaders to reduce cost, while increasing sales and investing for the future. He directed his computer executives to reduce overall information technology spending, while at the same time provisioning the company with a contemporary infrastructure and new computer applications that leveraged the business strategy. *The Digital Organization* explains how we accomplished this apparent contradiction in priorities.

The postmortem on the troubled financial information project distilled our mistakes down to three tenets that now guide AlliedSignal's deployment of new computer technology: (1) Install a technical infrastructure using aggressive corporate initiatives, (2) select the right technology, and (3) effectively manage computer people. It takes a comprehensive detailed strategy to integrate people, processes, organizational design, and computer systems into a seamless whole. When all these elements work together, the organization executes with what appears to be effortless ease. A truly digital organization complements the company's structure and processes with information technology that leverages skilled people.

We all want to energize and provision our organizations with the latest and best technology without challenging the national debt or chasing down too many blind alleys. Has AlliedSignal discovered a secret way of achieving this goal? I'm afraid not, but the company has made significant progress. In just a few years, it consolidated mainframes and servers, installed a global enterprise network, provisioned every employee with e-mail and groupware, upgraded site infrastructure, created an Electronic Data Interchange (EDI) processing center, adopted common systems across business units, and used Internet/World Wide Web technology for new and exciting applications. All of this was accomplished without ramping up expenditures. These achievements came from hard work, a good plan, and the application of standard management principles to computer technology.

Quite simply, if your plan makes sense, then you can achieve your objectives with surprising ease. The plans that most companies have for information technology cannot pass this basic test. A digital organization does not happen by accident. If you want information systems truly integrated with your people, processes, and organization, then you need a clear vision, a workable strategy, good plans, and determined execution. If there is a secret to AlliedSignal's success, it is that *it approaches computer technology with the same intensity and management style as every other aspect of its business.*

Tools of the Trade

Computers, networks, and applications are tools of modern business. Computers alone cannot drive profitability, but their absence makes it impossible for a company to be competitive. The information revolution truly has transformed business. Evidence of the explosive influence of computers can be seen in articles and whole sections dedicated to computers in *Forbes, Fortune, BusinessWeek,* and your local newspaper's business section. The shock wave of computer technology is powered by bright and ambitious entrepreneurs striving to be the next celebrity zillionaire on the cover of *Newsweek*. The computer industry epitomizes raw, aggressive capitalism, with rewards and recognition inciting innovation, invention, and ideas. The assimilation of all this new technology presents an enormous challenge to organizations. It seems impossible to keep up even though we spend huge chunks of money chasing the next piece of wizardry computer entrepreneurs bring to market.

The breadth and persuasiveness of computers rattles the senses. This all developed after World War II with electronic accounting machines (EAM), evolved to electronic data processing (EDP), progressed to management information systems (MIS), burgeoned into information technology (IT), and then exploded into myriad technologies indispensable to modern business. This technology outburst includes not only computers and networks, but voice communications, facsimile, videoconferencing, intelligent copiers, infrared devices, radio frequency units, and all sorts of specialty equipment that includes the ubiquitous microprocessor.

New technology allows us to improve traditional information systems, but more important, it brings to fruition whole new classes of applications. Software houses pump out office suites, work group applications, hands-off transactional processing, workflow management systems, electronic publishing, electronic data interchange, design tools, product management systems, automatic teller machines (ATMs), kiosks, virtual reality, and systems that allow customers to directly process their own transactions. The revolution has expanded beyond information to communications, imaging, document management, and process control. The high technology industries roll out new products in an unprecedented competitive furor. Running a business today means managing this outpouring of new computer technology—and making it all work.

Will Work for Food

Businesses and government want to take advantage of all the new technology, but they keep running into the same problem: legacy systems.

Each part of the organization does things its own way, using arthritic information systems. So management authorizes expensive projects to replace these old systems with new technology. When they don't work, time and money vanish without a trace. At a recent conference, someone sparked a lively discussion when she mentioned a $4 million failure in delivering a new computer system. After several rounds of "Can you top this?" the dubious honors went to an insurance company with a $30 million fiasco. Expensive system failures occur far too often and receive increasing play in the press.

Michael Ruttgers, CEO of EMC Corporation, claims that among his customers, management abandons one-third of new technology projects and three-quarters never fully deliver on their promised benefits. EMC, the largest supplier of off-line computer storage, sponsored a survey, to which 86 percent of the respondents expressed concern over their ability to use information technology to lower cost or increase revenue. The Standish Group, a Dennis, Massachusetts, consulting firm, says their analysis shows only 16 percent of information systems projects complete on time and on budget. The Standish Group confirms EMC's claim that companies cancel nearly one-third of new information technology projects.

Computer Associates CEO, Charles B. Wang, said, "You don't hear or read a lot about the failures, but some have been monumental."

Jeremy Frank, of GartnerGroup, chimes in, "Over half of all client/server projects fail."

Dr. Franklin Moss, President and CEO of Tivoli Systems, Inc., says, "What works in the lab, doesn't work in production."

Ambitious projects crash and burn on an increasingly frequent basis. Some are mismanaged, some stretch technology too far, some run out of money, and others are foolish from the start. Besides the systems that never get off the ground, a greater number get implemented, delivering far less than originally promised. All this frustrates managers, who lash out angrily at anyone within reach, especially the computer people. If this sounds familiar to you, then you've picked up the right book.

INTRODUCTION

Enterprising Solutions

Many huge successes, fortunately, offset the colossal failures. What makes the difference? Enterprise solutions provide the common thread of AlliedSignal's winning strategy. The business trend of adopting best practices relies on propagating systems and process across the entire enterprise. There is a basic law here: Enterprise solutions must be managed at the enterprise level. Corporate initiatives need to build a consistent infrastructure for enterprise applications to ride on. Without consistency, performance can be anemic, applications do not reach some corners of the enterprise, and costs increase as developers accommodate the junk strewn around the company.

The failure of big-ticket projects represents the biggest risk. To preclude this ugly scenario, corporations must initiate comprehensive upgrades of infrastructure and establish enterprise processes and standards for deploying computer technology. Sorry, there is no single silver bullet—slaying this beast will require firing off a great pile of silver BBs. Because technology moves fast, time is the greatest enemy; you need to catch up and then keep up.

The key to success is consistency, not technical eloquence. AlliedSignal accomplished this with far less money than some consultants will tell you is necessary. Speed and thrift require a disciplined approach with clear objectives. Since you don't have a lot of money to throw around, you need to reduce the cost of your existing computer systems to fund the new stuff. I'll describe the innovative ways AlliedSignal funded and rapidly deployed a contemporary infrastructure and corresponding applications on a global basis. A series of fast-paced programs built a solid infrastructure for the company and forced discipline on computer investments. AlliedSignal now demonstrates a keen ability to execute corporate initiatives to provide information technology to the entire enterprise.

Management Is Not a Spectator Sport

People in senior management positions tend to use one set of principles to manage their business, and an entirely different set when it comes to

7

computers. Why? Because the technology is complex and they've grown frustrated trying to make these machines do what they want. I'm going to explain how to get what you want from computers and how to manage this technology by using what you already know. Good management makes the difference between systems that hum and applications so anemic that they can't get out of their own way. Management frequently places too much reliance on some new piece of technology and forgets what it takes to make systems work. Throughout this book, I'll relate complicated computer issues back to general management principles you've already learned in the field.

My discussion will not get too complicated; I promise not to lose you in a morass of really technical data. I will explain what's working now, what's coming soon, and what you should avoid. Executives and managers do not need to become information technology experts. Most executives have never built a factory, designed a new product, defined a new chart-of-accounts, created a new advertising campaign, or written a legal brief. At least, they have not done all of them. Yet, they manage the professionals who do these things. You'll learn that it is equally important to manage your computer people.

The Digital Organization explains how AlliedSignal manages its technology to control spending and get the information it needs to compete. The book includes many suggestions on how to sift the good from the bad. These tips are important because the technologies discussed here will soon be obsolete, but you'll still be making computer technology decisions. Invention and innovation permeate the computer and related industries. You're going to learn about people who try unsuccessfully to slow down this change process. Some of them may even work for you. They're having a hard time managing the pace and don't even realize that doing so is their job. Confusion, conflict, and chaos reflect an energized and innovative industry that continues to change the way we live and work.

Organizations must learn to effectively deploy information technology. I've written this book for those responsible for delivering computer systems, as well as those who depend on other people's technological skills. I'm going to strip away the hype, explain what is going on, and guide you in directing information technology issues. I'll also explain how to leverage your previous investments in computers, so you don't need to start from scratch. This prescriptive book explains how to solve today's most pressing problem in business. I'm a practitioner, so everything I recommend I've actually accomplished in the real world.

Hang on, I'm going to move fast and I intend for this to be fun. By the time you finish this book, you will approach your management responsibilities for business technology with confidence and aggressiveness. You'll know how to organize your computer functions, judge the effectiveness of your business technology management, and assess the quality of strategic plans for technology deployment. We're going to demystify all the sacred cows and unmask the technical subterfuge that cost you money and hamper your ability to get the productivity you so desperately need in this no-holds-barred world economy.

A STRATAGEM FOR TURMOIL

Manage Your Strategy

An indefinable something is to be done, in a way nobody knows how, at a time nobody knows when, that will accomplish nobody knows what.

—THOMAS BRACKETT REED

Suddenly, your computers have become awfully ornery. Your old faithful workhorses have grown stiff with age while new technology was blossoming everywhere. Worse, your underlying technical infrastructure may be broken. There's more bad news—your job description says you must fix it. You know you need to shed this old stuff before you can invigorate your organization with new systems, but the task seems daunting. You don't have enough time, money, or able bodies. Additionally, someone has to hold things together while you simultaneously build a new world. You also need to pull this off without depleting the corporate treasury.

You need a plan—a good one. Most strategies for computer technology in the business world combine wishful thinking, dreadful economics, and naive simplicity. If your strategy offers up a clean-sweep agenda, costs more than Bill Gates's home, or relies on some flashy panacea, then you need to start thinking harder. A good strategy pursues a clear goal, drives tactical plans and decisions, conserves resources and, above all, provides the freedom to be opportunistic. Your plan should be thoughtful, concise, and known throughout the land. Many plans look good on paper, but don't survive the real world due to lack of commitment, poor execution, or inherent flaws.

I'm going to help you craft a strategy to build a digital organization that integrates your people, processes, and computer systems. Your plan must match your organization's culture and business strategy. Federal Express and Wal-Mart earned reputations for innovating new ways to leverage their businesses with high-tech computer systems. Each company designed unique solutions to business problems; neither copied anyone else. In observing executives, I have noticed that every successful leader has a style. The good ones develop their own style, resisting the temptation to copy someone they admire. The great ones hone their style to amplify their personality. It's the same thing with information technology: Match your business and amplify the organization's culture.

This will not be a technical discussion; important concepts are simple. I'm going to use AlliedSignal's experience to frame the discussion. We have accomplished an enormous amount of change in a short time without ramping up expenditures. AlliedSignal's success results from (1) using aggressive corporate technology initiatives to build a consistent infrastructure, (2) selecting the right technology, and (3) effectively managing computer people (Exhibit 1.1). These three basic requirements provide the framework for a strategy and for this book.

I can't tell you exactly what to do; you'll have to customize your own solution. Every organization is different and AlliedSignal's approach will not be a perfect fit. You'll also start from a different point of departure. Some of you have been grappling with this problem for years; others may

---

Exhibit 1.1
AlliedSignal's Strategic Framework

Use aggressive corporate technology initiatives.
Select the right technology.
Effectively manage computer people.

---

still be at the starting block. Whatever plan you end up with, current technology trends prescribe an initial series of mandatory steps for companies that want to take part in the accelerating information revolution.

A Glut of Opportunity

The opportunity for computers to inject adrenaline into your financial performance has never been better. Client/server and Internet/Web technologies make computers easy to use while expanding the applications computers can perform. Whether you want your customers to do business with you from home, combine all your customers' activity into a single application, use radio frequency (RF) devices to eliminate keyboards in your factory, create virtual design teams spread all over the globe, or increase aftermarket sales by mining customer records, modern computer technology can meet the challenge.

The problem is taking a conceptual design, assembling the parts, shaking out the bugs in the lab, and making it scale to the real world. It is a tough assignment—we've all experienced frustration with computer projects that have gone awry. The first to successfully deploy innovative applications using new technology can change the competitive positioning of an industry. This march of computer technology continues to unveil new ways to propel productivity, market share, and profits. You don't want to be left behind. The problem is leveraging the business strategy by creatively and economically deploying all this new technology.

The 64-Bit Question

The question that comes up at most computer seminars is how to align the computer technology strategy with the business. Exhibit 1.2 outlines the basic steps necessary to align computer technology with the business. Following these process steps consistently will insure that computer technology supports the business strategy.

I've never found the alignment issue difficult since I made a simple discovery that should have been obvious. After two years of being harangued by vice presidents to help them complete some unplanned

---

Exhibit 1.2
How to Align Computer Technology

1. Articulate and communicate a business strategy.
2. Set incentive-based objectives in alignment with the business strategy.
3. Computer technology leaders must understand the business strategy.
4. Computer technology initiatives must align with incentive-based objectives.
5. General management must invest time to assure the computer technology strategy is in alignment.
6. Use the same planning process for the business and computer technology.

---

project before year-end, I finally figured out that they had incentive objectives tied to these requests. I went to the president and requested that after he finalized objectives, he instruct his staff to have one-on-one meetings with me to reveal their commitments. After I made the rounds, I revised the computer technology initiatives to align with the executive's incentive objectives. It worked beautifully: There were no fourth-quarter panic drills, computer technology aligned with the business, and my relationship with other executives improved rapidly.

This only works if a business strategy exists and executive incentives align with this strategy. If neither of these is the case, then the severity of the alignment problem exceeds the prescriptive strength of the preceding advice. Contributing to the misalignment phenomenon are technology leaders who do not understand business. They become so enmeshed with technology, they never lift their head to see how it affects profits.

My first position with AlliedSignal included the assignment to turn around a lackluster applications group. No matter what I said or did, the applications manager just didn't understand my objectives. Finally, in frustration, I calculated the total payroll cost for his department and called him over.

I started the meeting by saying, "The payroll cost for your department runs slightly over four million dollars a year. Can you explain what your group accomplished in the last year to improve the company's profitability by more than four million dollars?"

He looked at me dumbfounded. "That's an unfair question."

I asked him why he thought it was unfair.

"Because we're not measured that way. We just do what the users ask, it's up to them to request systems that lower cost."

I could almost buy his argument, except that the "users" kept complaining that they couldn't use what finally came out of his applications group. The computer systems never met their operational requirements. I wanted more than a reactive, maintenance-oriented systems leader.

Technology leaders have an obligation to listen to their organization's business communications and understand how their discipline contributes to overall business success. On the other hand, general management frequently ignores their technology organization and assumes that somehow, some way they will get their act together. Executives who understand the link between their business and technology, spend a significant amount of time managing their computer investments.

A Winning Strategy

There are many ways to win. I classify strategies into two broad categories: doing the fundamentals better than your opponents, or outsmarting everyone with innovative approaches and products. AlliedSignal chose a hybrid somewhere toward the middle of these two extremes. Larry Bossidy even organized the company to support both strategies. He created an organization called Business Services and segregated the routine, repetitive operations from the businesses. Business Services includes Financial Operations, Human Resource Operations, Data Center Operations, Commodity Procurement, and Administrative Services. This shared service organization removes transactional activity from the business units and puts them under leadership driven to improve service, increase productivity, and reduce cost. While Business Services concentrates on the fundamentals, the business units search for innovative ways to compel customers to buy from AlliedSignal.

The Business Services organization only represents one aspect of Larry Bossidy's well-thought-out strategy for AlliedSignal. Other aspects include market predominance, restructured business units with delayered management, internal sales growth through customer focus, research and development concentrated in selected areas with a big potential, a single AlliedSignal culture, and taking successful businesses global. All this must be measured with appropriate metrics, and accomplished with speed within tight budgets that produce 6 percent productivity growth, compounded year over year. Quite a challenge. (Exhibit 1.3 illustrates AlliedSignal's business strategy.)

Sticking with the Game Plan

How does AlliedSignal align computer technology with the business? First, no one working for AlliedSignal has any excuse for not knowing the business strategy. Senior leadership communicates the strategy in speeches, articles, meetings, skip levels, and individual conversations. The consistency and repetitiveness get the message across.

Exhibit 1.3
AlliedSignal's Business Strategy

1. Market predominance—within top three.
2. Large business units for economies of scale and delayered management structure.
3. Shared services for repetitive transactional processing and centers of expertise.
4. Internal sales growth through customer focus and innovative products and services.
5. Concentrate research and development on high-potential technology.
6. Take successful businesses global to increase market.
7. A common AlliedSignal culture that is fact based and quality obsessed with satisfied customers and employees.

Accomplished with Speed and 6% Productivity

Next, Bossidy invests time on computer technology issues. Once a quarter, he chairs a technology review, quickly affirming when we are on track and not hesitating to redirect us when we miss the mark. The final meeting of the year deals with strategic planning and the first meeting of the year includes a Management Resource Review (an assessment of technology leadership and career development planning for high potentials). Bossidy meets with the Chief Information Officer (CIO), one on one, every other week to deal with issues too sensitive for an open meeting and to give planning guidance for the quarterly reviews. This doesn't count skip levels with technology professionals, technology discussions at senior leadership meetings, site visits to technology organizations, or Bossidy's frequent trips down the hall to catch the CIO on one subject or another.

The timing of the strategic planning session with Bossidy coincides with the business reviews and he insists on an identical format. This important autumn exercise ensures that technology aligns with the business. To start with, the technology organizations must commit to the same productivity goals. AlliedSignal's strategy calls for productivity gains greater than 6 percent, and technology wants to be part of the solution, not another problem the businesses must overcome.

Business unit consolidations, acquisitions, and divestitures require that the technical infrastructure be designed to accommodate the easy integration of new workloads and the stripping out of technology that supported a business or product line sold to another company.

Common solutions provide one of the enabling tactics to support AlliedSignal's strategy, especially the one-company directive. Common solutions require a consistent technical infrastructure so computer applications run smoothly across the entire enterprise. Since the target market includes the entire world, consistency must be maintained on a global basis.

To meet these objectives, AlliedSignal's computer technology strategy calls for provisioning the company with a contemporary infrastructure and applications that leverage the business, while keeping spending within historic and industry norms (Exhibit 1.4). In other words, spend

---

Exhibit 1.4
AlliedSignal's Strategy for Computer Technology

1. Common solutions (applications and processes) to the greatest extent that makes business sense.
2. Provision company with a worldwide consistent infrastructure.
3. Contemporary systems—balanced for cost, functionality, and speed of delivery.
4. Purchased solutions for routine activities.
5. Custom development concentrates on sales, product development, aftermarket support, and financial management.
6. Ability to integrate new business or separate the technology of a divested business.
7. New investments funded with cost reductions in existing systems.

Accomplished with Speed and 6% Productivity

---

smarter than the competition. To accomplish this requires reducing the cost of existing applications, redirecting spending to new technology, propagating common solutions, lowering the cost of new applications, providing an appropriate infrastructure, concentrating on applications that improve sales, and moving quickly. Beneath this strategic framework lies a consistent set of principles, strategies, and tactical plans. All this must be accomplished within budgets as strict as those imposed on the businesses.

All this effort has one objective: Deliver business applications to the workplace. AlliedSignal, like most companies, has moved increasingly to purchased software. Purchased software, along with common systems, gets answers out to the world of work faster and at less cost. AlliedSignal concentrates custom development on projects to support the field sales force, aftermarket sales, distribution, data warehousing and data mining, electronic commerce, and product development. Chapter 8 describes several of these efforts in greater detail.

After gaining market dominance, companies can get lulled into complacency by fat profits and soaring sales. Eventually, some hungry upstart comes along with a better mousetrap. If you want to make the industry's list of successful companies, keep your business focus on products and concentrate your best technology resources on product development, sales, and aftermarket support. You can only do so much, so direct your customized solutions to areas that will help you dominate your industry. It works for Federal Express, which continuously pursues technology to make it easier to ship with them instead of their competitors.

As mentioned earlier, the two strategy extremes are doing a superb job at the fundamentals or playing smart. Now, I've recommended being product obsessed. These approaches work together and their applicability depends on your type of business. Exhibit 1.5 shows the relationship between these two concepts.

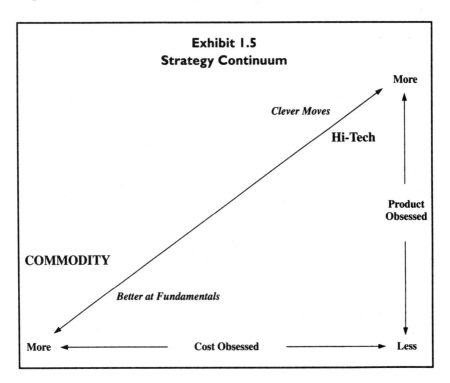

Exhibit 1.5
Strategy Continuum

More

Clever Moves

Hi-Tech

Product Obsessed

COMMODITY

Better at Fundamentals

More ← ———— **Cost Obsessed** ———— → Less

Commodity businesses must be extremely cost conscious, which mandates a focus on fundamentals. High-tech companies face less cost pressure, but must continuously bring new, exciting products to market to outflank their competition. Industries assumed to be mature frequently reignite competition with an innovation that suddenly differentiates their products from the competition. To achieve high margins, constantly challenge your product line, looking for ways to lure new customers, even at a premium price.

The Infrastructure Thing

I started this chapter saying your technical infrastructure may be broken. When a new technology application goes south, the guilty often cop a plea by fingering infrastructure as the real culprit. What is infrastructure? Good question. This trendy term has come to mean anything necessary to make computer applications work in the real world. Larry Bossidy once asked me that question, and I told him, "Infrastructure is anything left out of the project business plan." He did not enjoy my joke.

We came up with a better definition. Infrastructure, at AlliedSignal, includes (1) shared resources across multiple initiatives, (2) an investment that cannot be justified by a single project, and (3) required resources whose attributes are subject to strict standards (Exhibit 1.6). Infrastructure should be managed at the enterprise level to maintain

---

Exhibit 1.6
Infrastructure Definition

1. Shared resource across multiple initiatives.
2. Investment cannot be justified by a single project.
3. Resource attributes subject to strict standards.

---

consistency across the entire organization. This broad definition could include technologies, systems, processes, shared services, or naming schemes.

The most pressing infrastructure issue deals with networks and the surrounding paraphernalia necessary to make these things work on a predictable basis. All the new, exciting applications use network computing (e.g., client/server and Internet/Web technologies)—and most companies have anarchy reigning throughout their networks.

No one told you how much money you need to spend to upgrade all your personal computers, local area networks, premise wiring, servers, and wide area networks to make network computing perform as advertised. Diagnostic and production control systems are also woefully inadequate. When a network computing application breaks, it takes forever to find out where so someone can go out and fix it. No wonder we're having so much trouble making all this new technology work. It'll get better; it's just going to take time for developers to step back from the theoretical to the practical and to put infrastructure in place.

Caveat Emptor

There is an old story about the talking parakeet. A man, browsing in a pet store, spied a parakeet with a sign that claimed it could talk. Intrigued, he bought the bird, but once he got it home, it wouldn't talk. When the man returned to the pet store, the shopkeeper asked if he had a little mirror in the cage because the bird would not talk without one. So the man bought a little mirror. The parakeet still didn't talk. Next the pet store owner insisted the bird needed a little swing and then a little bell. Nothing worked. After repeated trips to the pet store and numerous additional purchases, the parakeet still refused to talk. One day the poor bird keeled over and fell to the bottom of the cage with its little feet facing straight up. Just before he died, the parakeet asked, "What, didn't they sell birdseed?"

You probably buy a lot of stuff trying to make large-scale network computing applications work and no doubt hope to find the requisite birdseed before your endeavor goes belly up.

Natural Law

How much backup technology will you need to implement before your enterprise applications speak fluently to all your people? Quite a bit, I'm afraid. Why? Because it's the law. If this surprises you, then you need to understand the Natural Laws of Computing. Here is the first one:

> **Law #1   Technology breakthroughs require a surrounding infrastructure.**

We build things to accommodate existing technology. When something dramatically new comes on the scene, it alters the attributes of everything that surrounds it and impels many changes to make the new technology function freely within its environment. The new digital video disks (DVD) for movies use a technology that blows away VCRs for quality, compactness, and longevity, but it will take years for this new media to send VCRs the way of eight-track tapes.

Obviously people must replace their VCR players, which won't happen until the entertainment companies convert a large selection of movies to the new format. To benefit fully from the new technology, consumers also will need to replace their television sets. The National Association of Broadcasters estimates the cost of replacing analog TVs for the digital variety will be at least $187 billion. Factories need to re-tool for a different consumer product and video production operations suddenly have obsolete equipment.

Now, think about video rental shops. They own the wrong racking to display DVDs and they need to figure out a way for customers to visually scan the smaller media. They're also stuck with far more floor space than they need. Since they can't get out of their leases, they need to find a way to profitably use the released capacity. Despite the cost, DVDs will inevitably replace VCRs. The transition, however, will take a disappointingly long time for people making major financial commitments to the new technology.

Also, these early DVDs can't record. This means consumers will need parallel equipment for a while. Disappointment will reach its peak when people discover they need to go through a second wave of equipment replacement when recording DVDs hit the market. They'll also wonder what to do with the hundreds of hours of Christmas, birthday, and wedding celebrations on videotape. Not to worry, conversion services will pop up as soon as the market reaches critical mass. Of course, this new infrastructure will all become obsolete as soon as someone invents an even more compact medium.

Throw Out the Old, Bring in the New

Maintaining, or improving, your competitive positioning requires you to deploy contemporary computer technology throughout your enterprise. A digital organization seamlessly communicates between employees, sites, business units, applications, customers, and suppliers with systems that empower processes and strengthen decision making. An unfortunate prerequisite for incorporating new life-sustaining business applications is changing-out a lot of your surrounding technical base that supports older technologies.

A digital organization requires a consistent technical architecture, strict adherence to standards, a full complement of network tools and utilities, centralized services, and applications deployed to the workplace that leverage the business strategy. All this must be accomplished without exceeding Warren Buffett's credit limit. It may sounds like a tall order, but it can be accomplished with a good strategy and adroit execution.

A Virtual Exercise

Strategic planning is tough work. One year, we had exhausted ourselves preparing for the annual strategy session with Larry Bossidy when we finally came to the last schedule. Everyone looked at it dumbfounded— it asked for our twenty-first-century ideas! Some corporate staffee thought this would provoke some good brainstorming, but we were already brain-dead.

Someone eventually suggested a virtual work environment at home. This got a couple of groans, and then someone said, "How about a virtual home environment at work?" That got it started. By the time we finished, we had listed a number of good ideas. The best idea was getting past the year 2000 with all our systems still running and pay still spewing out to our respective bank accounts.

In this chapter, I reviewed the highlights of AlliedSignal's strategy for business and computer technology. AlliedSignal's strategy may not be a perfect fit for your organization—you need to decide. Whatever you decide, avoid drafting a brilliant, but inert, plan, because you have failed to mesh your strategy with the culture and talent base of your organization. Your strategy may make sense, but sometimes you need to adjust the plan for the culture . . . or change the culture.

CHAPTER 2

TAKE THE INITIATIVE

Manage Your Initiatives

A man is known by the company he organizes.

—AMBROSE BIERCE

When Bossidy arrived at AlliedSignal in 1991, he set about identifying a project that had an immediate payback and would start changing the culture. The board of directors recruited Bossidy from General Electric because AlliedSignal limped from one income statement to another. No exciting new projects rolled out of the corporate chute. In fact, AlliedSignal was at least three different companies: Bendix, Garrett, and Allied Chemical. These predecessor companies roughly corresponded to the three business sectors: Aerospace, Automotive, and Engineered Materials. Each sector, and the business units, operated with a great deal of autonomy. This was about to change.

When I first went to work for AlliedSignal, I noticed that when two people met, they would immediately identify themselves as a Bendix or Garrett alumnus. The Allied Chemical people didn't mix at all. General Electric, on the other hand, centrally directed operations and had a strong company identity. Larry Bossidy decided that AlliedSignal's feudalism needed to end, and his mantra became "one company," but he was swimming against a heavy current. Data center consolidation was one of AlliedSignal's first initiative that took aim at changing its culture.

Violating the Prime Directive

When Bossidy searched for an initiative, he found only a halfhearted data center consolidation program. This initiative required each sector to reduce down to one or two centers each. This didn't make sense to Bossidy. Although not technical, he was pretty sure ones and zeros looked the same in aerospace, automotive, and engineered materials. He decided the company needed one center for all of North America. This larger program received board approval in September 1991.

At the time, I directed the Information Systems (I/S) function for one of the larger business units. My responsibilities included running a West Coast regional data center. I had already completed two consolidations, so someone mistakenly thought I must be an expert. I was an applications person who always had an excellent operations manager so I didn't need to deal with the hard side of the business. But when the offer came to manage the corporate program, I accepted because the opportunity to help build an organization from scratch doesn't come along very often.

To Boldly Go . . .

The lack of any people, organization, or facility intrigued me. Bossidy had directed a new facility be built and staffed with the very best people available from the existing centers and sector headquarters. He wanted the new center inside an AlliedSignal business so that security, accounting, and human resources services could be provided by established organizations. The corporate office selected Tempe, Arizona, because of available space, the low cost of living, and the area's designation as one of the safest spots on the planet from the threat of natural disaster.

I started with one person—myself—an attic, and a storeroom. In October 1991, I accepted the assignment to construct an unplanned center, recruit 119 people from a pool of nearly 300, transfer the largest existing center by May 1992, put operating practices in place, close all

---

Exhibit 2.1
Data Center Consolidation

North American Business Case (in millions)

Sites to be consolidated (11)	$58.2
Single data center	$47.4
Planned savings	$10.8
One-time costs	$16.0
Actual savings (consolidation only)	$11.5

Savings Increased after Reengineering

---

11 centers in 18 months, and deliver $11 million in savings to the bottom line. We beat every one of these objectives. Exhibit 2.1 shows the original business case for the data center consolidation program.

By Christmas, we had visited every site with a road show, selected the management team, and designed the new facility. Demolition began during Christmas break, with the start of construction scheduled for January. We worked out of temporary quarters, doubled and tripled up in cramped offices. By March, we selected the entire complement of 119 professionals and were moving with blazing speed.

Make It So

We overstaffed on managerial experience. The type of people who started the Computing Technology Center (CTC) had already run large organizations and reveled in making things happen. We made decisions at an unbelievable pace. No analysis—we just asked who had experience in a particular area, called them in, and trusted their snap judgment. I used to drive to work wondering who made all the decisions. I knew I wasn't making them, but things kept getting done. My operations manager showed up the first week of January, asking for direction. I showed him some blueprints and told him to build me a hardened data center for $5.9 million and have it fully functional by May, ready to support the largest business unit in AlliedSignal. One day he came to see me to settle a dispute with our landlord—the only decision I made after our first meeting involved overriding local rules on the size of the refrigerator in the break area.

The remaining staff followed type. The management team came for the excitement and they recruited people with the same attitude. The philosophy of recruiting the best flowed down effortlessly because highly competent people want competent people around them. Our mission was to run all the company's computer and network operations with less than half the people while simultaneously planning and shutting down data centers all across North America. Doing this with people who had never worked together, without operating procedures, management systems, or a tested data center presented an enormous challenge. I used to say we only had three things to get the job done; time, money, and talented people, and the first two were in very short supply. It was so much fun, I'm surprised they paid me.

Scotty, I Need Warp Speed Now!

We all thought we had signed up for an 18-month sprint, but the data center consolidation signaled the start of an endless race. Before we could finish this assignment, Bossidy unveiled additional layers of his strategy. AlliedSignal does not single-thread initiatives. Once the momentum gets going on one, another pops up. Exhibit 2.2 lists the initiatives AlliedSignal has used in the process of building a digital organization.

The first surprise came when the CTC assumed responsibility for the largest center in Europe. Bossidy wanted it done all over again on another continent. Despite our experience, this assignment was far from a piece of gâteau. We needed to take a French data center dedicated to AlliedSignal's largest European business unit and make it cross-sector and pan-European. We ran into significant cultural issues.

When we mapped out a strategy, our concerns included the French center's inexperience with consolidations, resistance from other Europeans, inadequate facilities and processes, tax issues, and the lack of true pan-European service providers. The resistance issue drew the most attention because other Europeans would be concerned about jobs

---

Exhibit 2.2
AlliedSignal's Computer Technology Initiatives

1. Consolidate worldwide mainframe computing.
2. Build a global enterprise network.
3. Implement e-mail and videoconferencing.
4. Standardize the desktop.
5. Internet, intranet, WWW, and groupware technologies.
6. World-class computing technology center.
7. Electronic commerce.
8. Drive common solutions.
9. Upgrade site infrastructure and consolidate servers.
10. Deploy innovative business applications.

---

leaving their respective countries and service not being provided by their own nationals. We immediately discounted our American practice of recruiting from the closed centers because we didn't need very many additional people and we understood the difficulty of getting someone to relocate across international borders.

The first objective was to gain the confidence of the French data center staff, not an easy task because I selected an American as Director of European Operations. I picked a retired Naval officer who had lived in six countries, but his French experience consisted of a short stint on one of their warships. I knew he could make a quick adjustment to France, but I really chose him because he had managed two data center consolidations and his style encouraged teamwork, respect for others' contributions, and an appreciation of different cultures. After completing a cultural immersion course, he quickly won over the French staff by listening to their suggestions and correcting some long-standing problems associated with their existing workload and facilities.

The rest of our strategy made sense to us but conflicted with the wishes of general management. We wanted to start with consolidating the other French centers to build experience and establish the processes necessary to support far-flung business operations. While we concentrated our consolidation efforts in France, we planned to accelerate our

networking efforts in other European countries to gain technical cred-
ibility and do something the businesses wanted done before we took
away their data centers. Our business management, however, wanted us
to start with a German center because it presented the greatest savings
potential. Consistent with the AlliedSignal culture, we accelerated the
schedule to win a compromise that allowed us to proceed along the lines
of our strategy.

Both the North American and European programs were enormous
successes. We made every schedule and beat the numbers. Savings continued to roll in as we standardized and simplified the disparity between
operations. More important, we built a cadre of expertise and management skill that became a tool for everything that followed.

Aged Beaujolais

While we busily shut down data centers, Bossidy consolidated Allied-
Signal's business units. In Aerospace alone, AlliedSignal went from 27
divisions to 5. All of a sudden, business units were geographically dispersed. Managers wanted to talk with other managers, who were now
across the country, and so did salespeople, production people, computer
people, and especially the accountants. Each of these professionals had
a personal computer and they wanted to share files and reports, but AlliedSignal's network only spoke mainframe.

We had built a brand-new facility and staffed it with the best
AlliedSignal had to offer, only to discover that we had caged dinosaurs.
At least that was the popular thinking. Our original mission called for
consolidating mainframe computing, but technology moved in a different direction. The exciting new applications used network computing
technology and they communicated with a different language.

The CTC received a new rush order; build a multilingual network.
There had been so many briefings on enterprise networks that company
executives thought they already had one, so the CTC was instructed to
connect 400 worldwide sites before sunset. Actually, they gave us a little more time, but this assignment also required breakneck speed. In

the first year, we connected over 100 sites representing 80 percent of the company population. This accomplishment required excruciating work in the areas of design, addressing, configuration, and testing.

An enterprise network uses router based technology that allows different types of computers, including personal computers, to talk over the same line. The much touted information superhighway, or Internet, uses this technology. In fact, an enterprise network is an intranet that connects to the outside world through the Internet. Done properly, an enterprise network allows anyone with a computer to communicate with anyone else in the world who is similarly connected. In 1993, this included many professionals; today it includes nearly everyone.

Neither Rain, nor Sleet, nor Snow

You need a vehicle to take advantage of a superhighway. How do you send a note or trade files with someone if you don't have an envelope or a delivery system? Besides, Bossidy didn't consolidate all those divisions for the benefit of the travel industry. Immediately after connecting a site to the enterprise network, another CTC team arrived to install electronic mail. Twenty-four thousand people received e-mail in the first year, and today everyone in AlliedSignal can send an electronic note to their CEO. Some, with more computer knowledge than language skills, have taken advantage of this electronic open door.

Although, the company had mainframe-based e-mail for over a decade, the ham-fisted mainframe variety looked so ugly that few people tried it. Everyone uses modern e-mail because personal computer technology makes it intuitive and stylish. At first, some people at AlliedSignal didn't treat e-mail as important. After Bossidy and the sector presidents started issuing instructions via e-mail, it flowed downhill in a hurry. Now, I can send a note to anyone, at any level, with assurance that they will read it and respond. Most people check their in-box several times a day.

Attachments provide the greatest benefit of PC-based e-mail. The ability to almost effortlessly attach a spreadsheet, document, or presentation makes these systems extremely valuable. Someone across the country can comment on a presentation during the same workday. Professionals can trade elaborate communications in minutes. E-mail, combined with a few groupware products, facilitates people dispersed all over the globe working together. AlliedSignal's work style was much enhanced by the universal availability of e-mail and related groupware.

Videoconferencing represents another technology that allows dispersed people to work together without boarding an airplane. Bossidy ordered every major site equipped with a state-of-the-art videoconferencing capability. The CTC provisioned every site across the globe with videoconferencing rooms and provided desktop video to select people who would otherwise monopolize these rooms. When will desktop video be provided with every personal computer? As soon as prices drop below management's line-of-sight.

Standardware

Something else came with e-mail. Standards for personal computer software were imposed on the entire company. There are good reasons to standardize the desktop, but e-mail makes it mandatory. People need to open their e-mail and read it. That means attachments must be created by applications available on every personal computer.

AlliedSignal selected a standard suite of desktop software and installed it with e-mail on every personal computer. People gave up their favorite software and learned a product that allowed enterprisewide communication.

Beware, enterprise solutions cause tough budget hits the first time through. When personal computer products come in slowly, the cost doesn't hit the controller's radar screen, but replace it all at once and you have the attention of everyone with numbers to meet.

Crowd Pleaser

Junk represents the biggest problem with e-mail. Electronic mail is a push system, which means anyone can blast out copies to everyone with the slightest interest or a higher pay grade. When an organization puts in an e-mail system, people start complaining that they can't find the important messages because of the volume of junk. Intranet web pages and other types of groupware answers this problem.

What is groupware? This buzzword sells software, so many products now carry the name. Groupware really represents anything that allows multiple people to interact with the same information. This includes bulletin boards, the Internet, and the ever popular Lotus Notes. (The Internet should be called mobware!) Home pages, chat rooms, and Usenet all represent forms of groupware. Web browsers make the Internet novice-friendly, and this exciting technology changes the way we access information. Chapter 4, Highways and Byways, explains the use of groupware, Web technology, the Internet, and intranet at AlliedSignal.

You want to invest in groupware as a way to avoid spending extra money because of e-mail. As a push system, e-mail sends out multiple copies of everything. With the popularity of large attachments, you'll soon find your enterprise network clogged with the same stuff being sent over and over again. Groupware, including intranet pages, use a pull approach. You create something, post it somewhere, and anyone interested can then pull down the information or add to the central copy. Using a pull system reduces junk mail and your network starts flowing more freely.

All-Star Cast

As the Computing Technology Center (CTC) raced through their initiatives, the department changed faster than the technology it managed. Yes, it got larger, but more important, the skills, structure, and mission kept evolving to meet demands. Every initiative described in this chapter was consciously designed to prepare AlliedSignal for the twenty-first century, except building the Computing Technology Center. The CTC happened by accident, or at least occurred as a by-product of other initiatives. Someone had to assume the onus for executing these programs, and the CTC kept getting the call. Exhibit 2.3 shows the steps AlliedSignal used to establish a computing technology center.

Exhibit 2.3
How to Set Up a Computing Technology Center

1. Select most qualified people.
2. Mix skill base between seasoned professionals and young high-potentials with experience in new technology.
3. Integrate technologies (mainframe, client/server, and Web).
4. Include technology assessment in mission.
5. Provision with appropriate tools and infrastructure.
6. Superior leadership.

Always Provide Clear and Consistent Direction

The Computing Technology Center started out life as a mainframe data center—just a collection of seasoned professionals with a will to get the job done. This group had been hand-selected; they knew their stuff, manhandled obstacles out of the way, got energized during a crisis, took pride in their accomplishments, and more than anything else, loved technology. Given the challenge of a new piece of technology, they devoured it.

The consultants say mainframers can't make it in the brave new world of network computing. They're wrong. They believe this because most computer organizations have accumulated deadwood over the years and even good people can get dragged down to the lowest common denominator. An organization like AlliedSignal's Computing Technology Center presents a great way to let the eagles flock together again. But you need the right formula. Start with seasoned professionals, strengthen the organization with fresh, young talent, provision them with the tools of the trade, give them clear, consistent direction, and then get out of their way. You'll be surprised. A few cautions, however; don't merely give them research assignments, don't create multiple organizations under the same roof, and don't waste your best resources under poor management.

Many companies have one or more enclaves of computer technicians with a vague charter to investigate emerging technology. If you have such a department, consider eliminating it. When I went to work for AlliedSignal, I found two technicians assigned to investigate new technology. I discovered everyone resented their freedom from responsibility, but I dismissed the complaints until someone pointed out these two employees had not recommended a single new product for adoption into our technology base. That got my attention. After an investigation proved this allegation correct; I asked to see the technicians' evaluation documentation. They had none.

I asked how they could dismiss something without documenting the tested release level with the corresponding deficiencies so we would know what to look for in future releases. They claimed they read all the trade journals and they would know when something finally made the grade. Their arrogance angered me, so I returned to my office and designed an assessment form. I returned and handed it to them and said they could test no new products until they documented everything they had investigated for the last year. Both quit within 30 days.

New computer technology abounds. Don't squander resources trying to sort out the eventual winners because most bright ideas never survive the harsh world of business. Technologists always search for perfect answers. While they investigate, an entrepreneur steals the market with an 80 percent solution. If you want to investigate emerging technology, then give clear directions to your lab technicians and hold them as accountable as the rest of your ogranization. Otherwise, give the job of assessing new products to your operations staff; they will focus on technology ready for mass deployment.

Another common mistake is putting more than one technology in the same organization, but never integrating them. When one group works on new technology, while the rest of the staff handles routine assignments, it builds animosity between the groups. Worse, you get no benefit from putting them together. The CTC originally took flack when it broke client/server into traditional disciplines and integrated them with existing mainframe groups. The operating system folks manage mainframes as well as UNIX and other popular client/server operating systems. The disk people handle every type of "flat, round, and spinning." The same holds true for database administration, production control, program products, networking, and the Internet. This approach produces economies of scale, interdiscipline skills, and keeps the new people from relearning old lessons.

An example the technically oriented will understand involved a client/server application that ignored the potential for record contention in their database design. The CTC technicians had been dealing with this problem for years and knew that client/server architecture didn't make this response-time bottleneck go away. By using some old-fashioned mainframe rules, these technicians helped make adjustments to the logical and physical design to avoid this problem.

Lead from Strength

Don't pull your best people together and then lead them with mediocre management. If you do, you just made things worse; your local I/S organizations lost the people propping them up and you got nothing in return. Select your very best management team to lead a computing technology center, even if it hurts the rest of the computer technology organization. Remember, if the person you selected for a mission critical-assignment was not crucial in their previous assignment, you picked the wrong person.

After you get things moving, replenish the remaining organization with new talent from outside. Business unit executives focused too much on the hardware. By removing their data center, they had no choice but to concentrate on what to do *with* computers, rather than *on* computers. In other words, AlliedSignal wanted the divisional computer organizations to support the business with an applications orientation. Some made the transition, others didn't. After we ripped data centers out of the business, about half of the I/S leadership ended up being replaced. The new executives' profiles included extensive applications experience and an ability to communicate with business leaders. It was not an easy time, but you cannot make major cultural changes without some unhappy dislocations.

Paperless Commerce

The original CTC included a group of six professionals called Network Applications. This small group transformed a center for mainframe computing into a center for everty type of computing technology. Their initial assignments included Electronic Data Interchange (EDI) and mainframe e-mail. From the beginning, this group's charter included bringing added value to a standard data center. This simple, but profound, concept anticipated the emergence of network computing. Network Applications forced the data center to adopt new technology and stimulated the whole organization to think creatively about the services it offered.

Getting an EDI processing center going became their first priority. Centralizing this service provided one-stop shopping for the business units. The group established trading partner relationships, mapped out the electronic commerce transactions, and set up network connections with everyone who owned more than a couple of tin cans tied together with a string. When a customer asked for EDI, AlliedSignal could respond in a fortnight. Once all our customers got linked, we went after the supplier base, putting over two thousand suppliers on EDI within one year. Electronic Commerce includes more than EDI, and Network Applications became involved with trading engineering data, aftermarket applications, the Internet, and systems that supported customer queries about products or services.

E-mail became the real impetus behind the transformation of the CTC. Network Applications drove the company to PC-based e-mail and the corresponding standardization of the desktop. This shifted the orientation away from large-scale, central processors to the desktop and networking. Enterprise networking, groupware, Internet/Web technology, and upgrading site infrastructure converted the center from data to computing technology. Including the Network Applications group in the original organization structure forced the CTC to grow and keep up with the rapid changes in the world of computing.

Getting Common

Everyone talks about common solutions; AlliedSignal does it. So let's discuss the importance of common solutions. It costs less to support fewer systems, and you want to reduce maintenance, or support costs, so you can fund new technology. If you do common systems right, you adopt "best practices" across the enterprise and propagate something that already works. Common solutions also allow you to leverage software licensing.

Exhibit 2.4 lists these advantages, plus another benefit not always recognized—when you transfer someone between operations, that person already knows the systems in their new home. This may not seem important, but it is. The transferred person hits the ground running, and assuming you selected a high achiever, you get the benefit of superior leadership immediately. When I managed application groups, I hated to hear about a new vice president coming to town, especially a VP of Manufacturing. They always submitted endless lists of changes they wanted in a hurry. They knew what had made them successful in the past, and they wanted the same indices, control points, and reporting systems as in their previous job. Common solutions puts an end to this phenomenon because when they arrive, they find what they had before.

Exhibit 2.4
The Advantages of Common Systems

1. Lower support cost.
2. Propagate "best practices."
3. Deploy working systems.
4. Leverage software licensing.
5. Eases moves between businesses.

Everybody agrees with the goal of common systems, but how do you get it done? Again, there is no simple answer. Exhibit 2.5 lists AlliedSignal's approach. Larry Bossidy firmly believes in common solutions, but knows it will not happen by edict. The first answer is Business Services. This organization groups common processes under a single executive. Common applications result automatically as the shared services organization redesigns the workflow and implements new computer systems.

Shared services can only address part of the problem. Corporate systems development represents another AlliedSignal technique. Generic applications (e.g., health, safety, environmental) get developed or purchased once at the corporate level and propagated to every business unit. As mentioned earlier, many I/S managers changed after data center consolidation; the newly hired executives came with no vested interest in the legacy they inherited, and had to be true believers in common applications to secure the job. Speed also drives common applications. If someone hands you a killer schedule you have no choice but to look around for an existing solution.

The final way AlliedSignal pushes common solutions is force of personality. Bossidy wants common solutions, so people know they need to have a damn good reason for using a unique process to do something that he considers routine. Through quarterly I/S reviews, Bossidy and

---

Exhibit 2.5
How to Achieve Common Systems

1. Combine routine transactional work into a single shared-service organization.
2. Acquire or develop corporatewide applications at the corporate level.
3. Use speed to encourage common solutions.
4. Drive purchased solutions—adjust processes, not software.
5. Build a common solutions culture.

   Use Unique Solutions Only When It Makes Business Sense

---

the CIO apply constant pressure on the business unit computer people to adopt common applications.

This doesn't mean everything should be common. Because I/S resources are scarce, Bossidy wants them applied to systems that deliver a business advantage. When a clear case can be made for uniqueness to steal a march on the competition, AlliedSignal executives quickly give a green light. The Aerospace Sector president instantly approved an innovative aftermarket application after seeing a rough prototype. This system answered a pressing business need to consolidate the information necessary to accelerate the sale of spares, overhauls, and replacement parts. Having resources available to jump on systems that provide a competitive advantage is the reward for using common solutions for mundane activities.

Switches and Routers and Hubs, Oh No!

Common applications drive you toward a major upgrade of infrastructure. If geographically dispersed business units share computer applications, then the highway these applications ride on must be consistent. The CTC first consolidated mainframes, then built an electronic superhighway between facilities, implemented e-mail to transport wares along the I-Way, standardized shrink-wrap software on the desktop, corralled servers into a controlled environment, and recently managed a major infrastructure upgrade that extended from the doorstep to the desktop. This latest assignment represented a crucial step in preparing AlliedSignal for enterprise client/server and Web applications.

A big telecommunications pipe comes into a facility and lots of devices direct the right stuff to every nook and cranny of the building. Network computing technology will not work unless this infrastructure is predictable, meaning it must be consistent, and change-out for this purpose strains budgets. Most companies set standards for new purchases and hope for the best. No good. You need to bite the bullet and attack this problem. When you do, take control of the entire network so "network computing" becomes a reality. The popularity of the Internet has built an impression that this simple technology has eliminated the need to refurbish infrastructure. Unfortunately, this is not true. The Internet, intranets, thin clients, and engineering workstations intensify the need for a solid infrastructure: Network traffic increases tremendously, speed and reliability become crucial, and network tools become indispensable. Network computing requires modern desktop equipment, with extensive amounts of real memory. The entire network and all its components should operate in a 32-bit environment. Gateways to legacy systems must be carefully structured to preclude unacceptable response times, excessive cost, and the proliferation of innumerable boutique solutions. Network components must be consistent, compatible, and contemporary to deliver service at a lower overall cost.

This subject is so important that Chapter 4 is devoted to explaining the AlliedSignal infrastructure upgrade and server consolidation program.

Seeking Sanction

How does AlliedSignal get approval for all these initiatives? The first step usually involves a rough cost/benefit analysis. If the numbers look good, or a compelling soft case exists, then an executive sponsor gets recruited. It helps if the executive is Larry Bossidy, but often the onus falls on one of his direct reports. Next, a full-fledged business case is prepared for presentation to the senior leadership of the company. All this sounds logical and straightforward, but in practice AlliedSignal can be just as messy as any other large organization.

I remember one presentation that, although eventually approved, started rather poorly.

This initiative finally gained approval in the usual way—extensive lobbying, a little bartering, some arm-twisting, and a lot of hard work to sell the project as the right thing to do for the company. It helped that savings commitments from previous initiatives had been met and the cost reduced for existing operations. (Funding can only come from the income statement or the existing cost structure. Since this hurdle stops many initiatives, it is the subject of Chapter 3.)

The difference between AlliedSignal and many other companies is the will to move the company forward and a culture that encourages people to lay proposals on the table for debate. Corporate technology initiatives are integral to a digital organization. To obtain the approval, authority, and funding for initiatives of this magnitude requires open access to the corridors of power and open debate by those who wield power. However, if food is served, remember to dodge any flying debris!

CHAPTER 3

NOBLESSE OBLIGE

Manage Your Budget

Just know your lines and don't bump into the furniture.

—SPENCER TRACY

Creative information systems, actually delivered to the workplace, can give you exponential improvements in productivity and increase sales. You may have noticed that such systems are rare. Most companies just plod along making the same mistakes as everyone else. Luckily, you only need to be as good as your competitors, and thankfully, they're as inept as you. Just think how successful you could be if you broke from the herd and did something truly innovative that changed the dynamics of your industry.

You already know where the opportunities lie—with new computer technologies. Launching network computing requires a complex refurbishing of the surrounding infrastructure, and you can't provision your organization with slick, new computer technology for free, so where do you get the money? This quandary drives the first step in the formation of a comprehensive strategy.

There is an estimated cumulative investment of $1.3 trillion in existing systems—a seemingly insurmountable barrier. Conference speakers use this startling figure to scare executives witless. Next, speakers remind the already shaky listeners that their old systems will go belly-up in the year 2000 (most old systems use a two-digit year that will confuse the year 2000 with 1900). So, our old systems are ugly and rigid, they'll break for good in a few short years, and the current corporate tight-fisted philosophy doesn't receive requests for eight-figure budgets gracefully. It's enough to drive you to early retirement.

Before you check your 401(k), let's try to reframe the issue properly. The amount of investment in existing systems reflects an interesting, but meaningless fact. You can't do anything about it, and it won't help you solve the problem. Ignore this piece of trivia. You need to concentrate on your current spending.

No Free Launch

Right now, you pay salaries, make lease payments, buy new software, maintain existing systems, and invest in capital projects. Add it up and you'll probably be surprised. Besides, in the real world, home from the enlightening seminar, it's all you've got. Not enough? Probably not, but you still need to start here. How do you get more from what you already spend? Exhibit 3.1 gives you a sneak preview. This chapter includes a brief description of each action; later chapters will flesh out the details.

Before you can put together a coherent plan, you need to stop wasting precious money chasing after magic elixirs. I'm going to explain how

---

Exhibit 3.1
Funding the New Technology

Drive down legacy system cost.

Manage on an enterprise basis.

Lay down a consistent infrastructure.

Preserve existing investments.

Concentrate on applications that provide more money.

Manage existing resources, including computer people.

---

to get this job done, but first I need to focus your attention on a contributing cause of your frustrations—you may have exactly the computer systems you deserve. Scary thought? It's easy to blame your Information Systems staff, but before you fire your computer people, look in the mirror. How do you contribute to the problem? Because you want quick, simple solutions, you accept vague promises to overcome all the limitations of the past by using a new technology.

Right now is an exceptionally precarious time. An influx of new technologies promises to change the way we use computers. Just look at a few of these buzz concepts: client/server, Internet, intranet, World Wide Web, open systems, enterprise e-mail, asynchronous transfer mode, virtual reality, outsourcing, data center consolidation, data warehouses, data mining, object-oriented everything, rapid applications development, enterprise integration, and on and on.

As advertised, new technology provides opportunities to run our businesses and organizations in new and exciting ways. These opportunities are on hold, however, while your computer professionals learn to build new systems with speed and dexterity. Applications design and development still costs too much and takes too long.

You need to avoid the temptation to grab the first thing that comes along that promises to make this problem go away. When someone waltzes in and promises you exactly what you want in record time and at a steep discount, you may regret being gullible and going for it. Computers can be made to do what you want, but there are a few prerequisites: you must know what you want; your expectations must be realistic; and you need to manage information technology like everything else, with a practical strategy.

Controlling the Burn Rate

To drive the cost out of doing business today, you must examine your portfolio of existing computer systems, which consumes most of the technology budget. We call these old applications "legacy systems," a term that tends to disparage anything that doesn't run on a personal computer. Legacy systems run on mainframes, architected by IBM, or mid-ranges from a host of companies. Maintaining these applications chews up the lion's share of most technology budgets. Most people look at this problem and throw up their hands—don't, it's your first source of funds.

Consolidating computing and telecommunications provides a quick way to reduce the cost of existing systems. This works for both legacy systems and the new technology. Most astute companies have already consolidated their mainframe computing into a single center. If you haven't, then you are behind the power curve. Companies that aggressively attack cost, consolidate their servers and mid-ranges. This may not seem as obvious, but recent technology developments make it possible to run these smaller boxes at far less cost when corralled into a single location.

How much money can be saved? It depends on the size and number of centers you have today. AlliedSignal took 40 percent of the cost out of data center operations by consolidating 11 mainframe sites into one. Server consolidation savings are not as dramatic, however, centralizing this operations improves security and disaster recover, while positioning you to take advantage of emerging technology to manage large server farms. Data center consolidation can provide funding for other initiatives, so it should be the first step in your strategy. It can be done fast, savings are predictable, and users don't see any disruption to their operations.

Reducing application maintenance provides another way to decrease legacy costs. I do not bemoan spending resources on maintenance. I've managed applications for too long to believe maintenance adds no value. Maintenance really represents changes necessary to keep computer systems current with business practices. Most people associate the term maintenance with keeping things working. Applications maintenance includes everything from small changes to support new accounting rules, to major modifications to further integrate business processes. Without these changes, legacy systems will soon hobble your organization.

Despite the usefulness of maintenance, the costs should be wrestled down. Maintenance may be valuable, but everything is relative. By arbitrarily racheting down the maintenance budget, you'll be squeezing out the less worthy changes and freeing up resources to work on new technology. Just remember you're making a budget decision, not a value statement.

Intruder Alert on the Enterprise

Everything I explain in this book works much better if managed on an enterprise basis. Exhibit 3.2 lists the reasons an enterprise approach delivers superior results.

Mission-critical applications require reliable delivery of service. Networks constructed without good planning prove unreliable, but more important to our current discussion, they cost big money. Most local area networks (LANs) are held together with duct tape, baling wire, and chewing gum. You have a lot of people out there trying to keep these things working and you need to redirect them toward building new applications. To control cost and put order into the current chaos requires a consistent technical architecture across the entire enterprise.

You can't fix this mess by allowing each business unit to muddle their way through to their own solution: not if you want enterprise computer applications. There's a basic law here:

Law #2   Enterprise solutions must be managed on an enterprise basis.

Despite its simplicity, this concept often doesn't find a receptive audience. For a successful digital organization, however, you need to make it happen. Besides releasing people for more productive work, enterprise

---

Exhibit 3.2
Enterprise Solutions

Reduce people cost.

Leverage purchasing decisions.

Preserve existing investments.

Force consistency with business plans.

---

solutions allow you to leverage purchasing decisions, increase the preservation of existing investments, and force consistency with business plans. Pick a standard and then ask your buyers to negotiate steep discounts with volume purchase agreements.

You want to preserve your installed base to the greatest extent possible and propagate the best-in-class across the rest of the enterprise. And last, if you want your information technology to propel your business strategy, then you need to direct it from the top. To realize these benefits requires an enterprise perspective and corporate clout. Point solutions nickel-and-dime you to death.

Hobgoblin of Small Minds

Consistency is the secret to lower costs. This axiom holds true for your technical infrastructure as well as for the repetitive processes in your organization. Whether you know it or not, you already spend a lot of money on networking. Somewhere out there, people buy all kinds of gadgets and gizmos in an attempt to piece together a solution. You need to measure this spending and take control.

After data center consolidation, the physical infrastructure provides the next best opportunity to get more from your spending. Start with the wide area network. Telecommunications presents an excellent way to take existing cost out of your technology budget. You can leverage pricing, and recent developments allow your technical people to rationalize networks by combining different types of traffic. (Voice and different types of computer traffic can now be transmitted over the same network.) The next step should be to tackle the local area networks inside your facilities. This win/win strategy reduces cost and builds a manageable consistency into your infrastructure. Your computer technicians will argue incessantly about the nuances of one brand versus another, but in the end it makes little difference. You just need to get it done—the sooner, the better.

A Penny Saved Is a Penny Earned

You can't replace everything you have today. You can't afford it. So start looking around and figure out what can be salvaged. Remember, everything you have is not junk. When AlliedSignal looked at its infrastructure, we compromised on some elements because of the installed base. If some element didn't totally meet our technical requirements, but we already owned a large amount, then we asked ourselves if we really needed to change-out this device. Would the expenditure really represent the best use of our funds? If not, we replicated this also-ran throughout the rest of the company. We spent some money buying equipment our technicians didn't rate as best-in-class, but we controlled our overall spending and achieved the consistency we desired.

Likewise, you can't just toss out all your legacy systems for spiffy, new applications. You're not going crack that $1.3 trillion nut. Besides, much of this old technology works just fine. Instead of replacing these systems, you need to do some work around the edges. Approach your legacy systems like you would an aging, but well-built house. Fix the plumbing, knock out walls to improve traffic flow, update the interior, and apply a fresh coat of paint to everything in sight.

Examine your installed infrastructure and preserve any existing investment that makes sense. Change occurs fast in the computer field. Don't fret over how far behind you may be; no one can afford to stay current all the time. Guess what? Your competitors are in the same bind.

The Source of Milk and Honey

The quick-paced introduction of new products makes the computer industry exciting; the rapid obsolescence of yesterday's purchase makes it career threatening. Despite all the cost savings ideas I'll present, it will still cost more to deploy new technology than you can squeeze out of existing spending. Alas, your customers provide your only source of money. The brutal fact of life is that the technology budget gets set as a percentage of revenue for your enterprise. So, when you evaluate spending priorities, look for investment with the greatest impact on sales. If you need more money, then work to raise revenues. AlliedSignal concentrated their new application spending in this area because the strategy fit with Bossidy's priority on internal growth. Customer service, aftermarket, and electronic data interchange (EDI) provide quick-hit opportunities.

Use What You Already Have

Whenever someone asks Larry Bossidy for money, he asks how much they spend on defects and scrap. He reminds them that it's their decision; they can spend their money fixing errors or use it to build the business. The same thing applies for information technology. Bossidy forces an enterprise spending analysis by function. Across all of AlliedSignal, between $300 and $400 million gets spent annually on technology. That's a lot of money. After you tout it up, you need to slice and dice it to figure out where it gets consumed. Some decisions can then be made to optimize spending and redirect resources.

When examining your current resource allocation, don't forget to look at people. "Plug and play" is a marketing fiction. You can't just buy a piece of software or hardware and assume that it will leap from the delivery dock into your integrated, state-of-the-art systems. Someone needs to put the pieces in place, and the total picture probably looks like a three-dimensional jigsaw puzzle—it requires experienced and skilled people. Your best people may not be applied to your biggest payback projects. Shuffle them around. Again, this takes an enterprise perspective because your business units will husband whatever talent they have to apply to their own priorities.

Controversial? You bet. AlliedSignal provides cover for this program with a cross-functional career development program. Promotions became scarce for employees who don't participate, so high-potential people make sure their names surface when evaluations are made. This allows assigning skilled resources to the important corporate initiatives.

Prime the Pump

Will the money you obtain from existing sources be enough? I don't know. It depends on where you are, where you need to be, and how fast you need to get there. I can tell you this: Additional money comes easier once you've demonstrated success and proved you're doing everything possible with what you already have under your control. The rich do get richer and winners keep on winning.

Okay, say you have the money, not enough to do everything, but enough to get started. Now, how do you pick the best opportunities to demonstrate what can be done with all this glitzy technology? New products come to market every day. Most of it solves one problem or another. The computer industry will soon solve the bigger issues. However, we need to find ways to get real solutions out to the workplace at a faster rate.

When researchers come up with something new, you know it takes hard work by scores of people with different skills to actually take the product to market. The same thing is true with information systems. Computer professionals must be held as accountable as your design engineers who routinely accept the challenge to bring expensive development projects in on time and within budget.

Sorting the Wheat from the Chaff

With proper management, the selection of technology investments is not difficult. Of course, there have been high-profile failures in implementing new computer systems, but progress never comes without pain. Sometimes you just have to jump right in and take your lumps. Obviously, you can't charge after every technical advance with an open wallet and unlimited patience. Everything does not eventually succeed. Besides, no one has enough money to chase everything on the horizon. You need to pick and choose.

You may need to be leading edge for information technologies strategic to your business. For other things, you may wait until the early adopters bleed, recover from their wounds, explore new directions, and eventually blaze a trail for you. Finally, you may decide to wait on other things until they have been so well established that you can pick a plug-and-play solution. It's your choice—if you don't just let it happen. Most companies don't make their information technology decisions systematically. They do it haphazardly, by luck of the draw, or they just grease the squeaky wheel.

How should you make investment decisions on all this new stuff? These are big ticket items and the technology can be intimidating. You don't understand it, you don't know if it will work, and the delivery record does not build confidence. But one day, someone whispers in your ear that you must do this because:

- Our competitors are doing it.
- It's strategic and it will save us piles of money.
- Objectives and commitments cannot be met without it.
- Everything we have got is obsolete, inflexible, and worthless.
- All the above.

Several people probably bend your ear on their pet projects. How to choose? The technology seems all over the map, each looks promising, and all the advocates promise untold riches just beyond the technology barrier. There must be a way to sift through these projects to find the real gems. Earlier, I promised to show you how to manage IT issues by using what you already know. Here's the first tip; you select information technology projects the same way you do capital planning.

A Capital Idea

Capital planning starts with an annual exercise where everyone puts together a wish list of everything they believe will improve the bottom line. Items are selected based on a cursory analysis or known infrastructure needs. Submitted investments may include machine tools, test and lab equipment, a new roof for the plant, paving the parking lot, items to comply with regulatory requirements, computers, or even an entirely new factory. After the compiled list is toted up, it usually exceeds the available budget, to no one's surprise.

Now the fun really starts. Things that must be done get priority and the remaining items survive only after a painstaking evaluation. Presentations on the bigger items assure that the finalists pass a peer review and support the organization's strategy and short-term goals. Outsiders question the technical risks and the sponsor must explain how the new approach will blend with existing processes and support long-term directions. Eventually the weak entries get weeded out and a capital plan is submitted for higher approval.

Another surprise; it comes right back down with a request to cut it further. Arguments are strengthened on why each item is crucial. Minor bickering and infighting develop as each department head competes for a fair share. In the end, the top executive has to weigh the merits of each request and make some final decisions. It's a messy, anxiety-laden process, but it works because it includes extensive peer review, scrutinized technical feasibility, evaluation of infrastructure issues, identification of prerequisites, plans for integration with existing operations—and the top executive stays engaged.

Now compare this process to how you make your information technology decisions. Both capital investments and development of a major computer systems cost big money and have long term benefits to the organization. Both have the potential of positioning you to execute your business strategy. Failure of either adversely affects your financial performance. Decisions on both should go through the same rigorous process.

You will avoid the most serious mistakes if you follow the capital planning model. You can be in control. *You need a little knowledge of the technical stuff, you must be realistic in your expectations, selection must be based on strategic and tactical needs that fit the specific requirement, you need to know the maturity level of the technology being proposed, you need to understand the biases of the proponents, and you need to select and plan in a disciplined manner.*

System Limitations

If you do successfully build a digital organization, don't expect to resolve all your business problems. If your processes or management need attention, then automation does not provide a solution. I've frequently witnessed overreliance on a new system to solve nontechnical business problems. Misplaced faith in technology frequently provides a ready-made excuse for delaying other actions required to improve performance.

I learned this lesson early in my career when I worked for a large air-craft manufacturer. The company's major military contract was nearly a year behind schedule and chaos reigned throughout manufacturing. Because part shortage had nearly shut down the assembly operations, the director of production control decided to eliminate our Material Requirements Planning (MRP) system and to allow the assembly departments to directly schedule the fabrication departments. Essentially, we switched from an MRP's push approach to a pull system. This was actually a superior way to schedule the factory, invented out of necessity before the Japanese just-in-time scheduling concept became popular in American factories.

As the systems analyst assigned to production control, my job was to develop the new priority scheme and present it to the 10 assembly managers. After completing the system, I went to each of the assembly managers explaining how the system would work. The forward fuselage manager nearly cried as he explained he needed all the help he could get. My problem was the nacelle manager. Despite numerous attempts, I was never able to find him.

Finally, his secretary confided that if I really needed to talk to him, he could be found at a desk located amidst the manufacturing engineers. She was right. He sat at a standard military-issue desk in the large bullpen, ignoring his office and private secretary. I sat down next to him and explained the new factory scheduling system.

He listened for a while, then picked up the report and threw it on the floor, shouting, "Any manager that doesn't know what parts he needs, shouldn't be in the job."

I was more than a little intimidated, but I managed to pick up the report and put it back on his desk. Prior to making a quick exit, I said, "It doesn't matter whether you like the system or not. Starting with next Tuesday's shortage meeting, this is the only way you'll be able to schedule parts for your department."

I hadn't realized that slowing down the progress on the nacelles motivated the design of the system, which accounted for the nacelle manager's anger. The forward fuselage was far behind schedule, while the nacelles sprinted ahead of everyone else. Unfortunately, the fuselage mated first, and the nacelles, which held the engines, got hung off the wings just before the plane left the hanger. The hidden agenda for the new scheduling system was to move priority away from the nacelles to the fuselage.

I received a surprise at the next shortage meeting. The only person to have properly prioritized his shortages was the nacelle manager. He also thoroughly understood the summaries and statistics at the end of the report that reflected the status of his assembly area. No one else really understood the system, nor had they taken the time to properly set up the control mechanisms. Three months later, the nacelles were further ahead of schedule, and the fuselage continued to have problems.

About this time, I accepted a transfer to a midwest division, leaving this mess for others to sort out. When I returned several years later, the subject of problems with the early stages of the program came up. I asked, "Did they ever resolve the schedule problem between the fuse-lage and nacelles."

"Oh yeah," someone responded, "they switched the managers be-tween the two assembly sections."

Once you finally get the systems you need, the divergence between your best and poorest performing managers and business units will grow. I've seen this over and over again. You can't crutch bad managers with a good system, but good managers use everything within reach. Good managers leverage information systems, and poor managers tend to blindly rely on someone or something else to make up for their short-falls. You also need to be aware that good managers tend to get good systems, while poor managers tend to blame the computer for not being able to get their job done.

The same phenomenon holds for processes. Automating an ineffec-tive process further ingrains the inefficiency into the organization. I'm not suggesting that you delay pursuing a digital organization until you've addressed all your other issues. Technology and your competi-tion move too fast for you to wait. My recommendation is to improve your organizational design, processes, and people simultaneously with the implementation of new computer systems. A balanced, multithrust approach increases the benefits of each of these four pillars of organizational performance.

Now let's start crafting a strategy for upgrading your infrastructure. Examining AlliedSignal's approach will give you tips on how to customize your own plan. Put on a hard hat because a lot of brickbat gets thrown as you work toward consensus!

CHAPTER 4

HIGHWAYS AND BYWAYS
Manage Your Infrastructure

*We are in great haste to construct a magnetic telegraph
from Maine to Texas; but Maine and Texas, it may be,
have nothing important to communicate.*

—HENRY DAVID THOREAU, 1854

The Information Superhighway (I-Way), cyberspace, and World Wide
Web all create the image of a high-speed, interconnected thoroughfare
for computers. If you've traveled on the I-Way, then you know these Internet nicknames represent hyperbole. The Internet more closely re-
sembles the road system Dwight D. Eisenhower found after returning
from World War II. Worse, getting to your destination on time and intact
is problematic because of the absence of road signs, traffic laws, or security. The Internet congestion keeps getting worse as Web browsers attract new users and encourage experimentations.

The graphical user interface (GUI, pronounced "gooey") that came
with personal computers made them easy to use. Web browsers and
Home Pages do the same for the Internet eliminating the need for a
computer science degree. All this traffic clogs the system, so network
computing still requires you to subscribe to private toll roads (i.e., a
major carrier's leased capacity)—and this will not soon change. To

stretch an analogy, even if the I-Way was a secure, six-lane divided high-way, you would still need to build your own off-ramps and surface streets. If you want to facilitate network computing, with easy traffic flow within and between your offices and plants, then you need to stan-dardize and upgrade your network infrastructure.

Some Assembly Required

Networking presents the biggest hurdle to building a digital organiza-tion. Internet/Web and client/server technologies drive an increasing dependency on networks. Actually, traditional mainframe computing started the network trend when companies began consolidating data centers. When you stash the computers away from the people trying to use them, then you need fiber, wire, or radio waves to connect them. Mainframes use a one-to-many architecture with hundreds or thousands of "dumb terminals" attached to a central computer. This is a relatively simple design and the maturity of the technology makes mainframe net-works reliable. Today, most people use a personal computer to attach to a mainframe, but these intelligent devices shed IQ when talking to big iron.

Client/server brought complexity to networks. No single supplier, like IBM, provides all the necessary gear with a single, consistent architec-ture. Client/server systems, like mainframes, use a hierarchical archi-tecture. But, instead of dumb terminals, an intelligent client handles part of the processing. This means heavy, uneven traffic and a lot of network coordination between the client and the various servers. No standard architecture exists, so many manufacturers and vendors supply parts. The Internet, with its peer-to-peer architecture, takes network com-plexity to a whole new level because of its many-to-many connectivity.

On my fifteenth birthday, my mother gave me a spark plug as a gag gift. She said it was a car on the installment plan. I can imagine the kind of car I could have built picking up parts here and there over an ex-tended period. But that's the kind of networks most organizations use to link up their personal computers. When they run high-speed, mission-critical applications on these things, it's like bringing a jalopy built by car-crazed kids to a NASCAR race—you know it will break, crash, or snarl traffic after one or two laps.

Running on Empty

For decades, I believed that the demand for computing power was insatiable. Today, I can place enough computing power anywhere necessary to satisfy the most demanding customer. The problem has shifted to bandwidth. Computing power—combined with network applications, graphics, and multimedia—has outstripped most networks' ability to handle the traffic with any kind of reasonable speed. As mission-critical applications migrate to network computing, the relatively poor reliability of most companies' infrastructure becomes of paramount importance. The level of expectation has also increased as people want networks to be more reliable than a dial tone.

Exhibit 4.1 lists the reasons to upgrade your infrastructure. Because of the cost, these programs don't get a quick nod. You need to work up a justification that appeals to your own corporate culture. I'll go through the major points in the hope that several items will strike a resonate cord.

Exhibit 4.1
Reasons to Fix Your Infrastructure

Improve network speed.

Increase reliability.

Deliver applications to workplace.

Eliminate programming for multiple environments.

Decrease mean-time-to-repair.

Reduce desktop support costs.

Support geographically dispersed work teams.

Accurate asset management.

A Prerequisite for New Technology

Speeding Down the Off-Ramp

Modern applications chew up disk space and bandwidth. The first is easy to solve, the second a bear. Even if you subscribe to huge networks between your facilities, you probably encounter a choke point with your existing premise wiring and local area network infrastructure. The technology exists to deliver Niagara-like flows to the desktop, but your only choices include waiting for wireless to develop in the next few years, or pulling new wiring throughout the building. Rewiring involves heavy costs, disrupts operations, and may carry numerous health and safety risks. Even so, most people can't wait and need to attack this immediately. If you find yourself in this predicament, then I advise installing more bandwidth than you think you'll ever need—you really don't want to do this again on your watch.

Dependable Transportation

Computers—both mainframes and the stand-alone personal variety—possess a well-deserved reputation for reliability. The problem comes from tying personal computers together and then implementing mission-critical network applications. If people can't get their work done because of network problems, the decibel level reaches the corner office. When someone's job performance becomes jeopardized, it does no good to explain the increased complexity of network computing. You need to tackle the problem and fix your network infrastructure.

A Car in Every Garage

Delivery of network applications to every corner of the enterprise presents another difficult problem, especially to remote locations and factories. These applications (e.g., serve-yourself human resource systems; health, safety, and environmental applications; client/server based material requirements planning; and applications designed for your sales force and traveling executives), all require a contemporary infrastructure to provide easy access to the corporate network. Frequently, the justification for these enterprise applications assumes universal access, and only later does the project team realize that network incompatibilities disenfranchise a significant portion of the population.

Speed Bumps

Development teams go to extra effort to accommodate disparate hardware and software strewn about in order to provide everyone access. On one AlliedSignal endeavor, a postmortem identified that 30 percent of the development hours were dedicated to special programming for nonstandard equipment in the field. Application upgrades also become dicey if network configurations change on the whim of a local techie. We had to back out of a few client/server application upgrades due to unanticipated network tampering at remote sites.

Collision Avoidance

Major opportunities exist to reduce the cost of supporting networking computing, but these easily quantified savings don't compare to the productivity of giving people fast access every time they want it. A fast mean-time-to-repair provides the only way to achieve 99 percent (or better) uptime. Why? Because things break—it's another law.

Law #3   Things break!

Studies show that over 80 percent of repair time involves isolating the component or glitch causing the problem. Once technicians figure out what broke, they usually can fix it without delay. Industrial strength networks require automated diagnostic tools that isolate problems fast so you can execute a repair almost immediately.

Road Work

Since most corporations want initiatives justified with hard currency, many infrastructure upgrade programs seek approval based on lower support costs, rather than on a qualitative assessment of improved performance. Merely centralizing Help Desks offers saving, but the big numbers come with diagnostic tools, electronic software distribution, and LAN backup, as illustrated in Exhibit 4.2. When the central Help Desk can detect a problem and fix it before anyone notices, then you save money and reduce stress on your already overwrought employees. Studies show that electronic software distribution saves the most money

Exhibit 4.2
Lower Support Cost

Centralize Help Desk.

Automated central diagnostics.

Perform some corrective actions centrally.

Electronically distribute software.

Centralized server backup.

Real-time training.

Strict change control.

by eliminating the deskside visits of countless technicians to download the latest software release. Actually, getting everyone started with the same version bright and early Monday morning provides more benefit than the lower labor costs. Similarly, doing centralized LAN backup protects valuable data, as well as saves money.

Don't wait for Web technology and network computers (NC) to eliminate the need for electronic software distribution. Even if you adopt an exclusive intranet architecture, it will take years for you to use the new Web languages to replace your existing application portfolio. In the meantime, your technicians and your budget need relief from all those deskside visits.

Many people bemoan the evolution of help desks to real-time training. Users want to call someone to get them out of a bind rather than read manual's or context-sensitive help screens. Although this puts pressure on Help Desks, training everyone on every feature does get expensive.

The biggest cost savings don't come from all this electronic wizardry, but from applying strict change control on your personal computer technicians. Outages happen only because something broke or something changed. I've already mentioned that modern diagnostic tools help cut repair time, but the truth is, things don't break nearly as much as you may have been led to believe. Most outages result from

someone changing something: hardware, software, firmware, configurations, addressing, processes, and so on.

Law #4   Change causes the most downtime.

If you want a quick-hit improvement in reliability, make your technician *document their changes in advance, test them thoroughly, prepare
back-out plans, and force them to schedule changes deep into weekend
darkness.* Control everything from a central change management system so your help desk can shoot problems. This will be hard on your
technical staff, but it will significantly improve reliability.

Traffic Patterns

The trend toward decentralized work teams, globalization, and best
practices requires people to communicate with voice, pictures, and data.
Sophisticated voice communication, videoconferencing, groupware,
e-mail, telecommuting, Web applications, and file sharing all depend on
networking. If your business plans include leveraging people scattered
all over the planet, then deploying a contemporary network infrastructure becomes a prerequisite of your strategy.

Vehicle Inspection

Asset management provides another valuable, but underrated benefit of
upgrading your infrastructure. Like many companies today, AlliedSignal
couldn't get a handle on the number of workstations it owned, the type
and configuration of its network black boxes, or the software portfolios
loaded into its servers and personal computers; nor could we get an accurate measurement of desktop spending. (When we implemented e-mail,
counting our addresses gave us our first hint at how fast these desktop
marvels breed.)

A major feature of a contemporary network infrastructure includes
the ability to electronically query every nook and cranny of the network to build an accurate inventory of these expensive assets. Modern
tools that take an electronic census of your network can also catalog the
internal configuration of personal computers. Combine this capability

with a change control system that includes asset management and you can measure capital spending, enterprise depreciation, and compliance to standards.

Dream Machine

In *Future Shock* (1970), Alvin Toffler said, "Each new machine or technique, in a sense, changes all existing machines and techniques, by permitting us to put them together into new combinations. The number of possible combinations rises exponentially as the number of new machines or techniques rises arithmetically. Indeed, each new combination may, itself, be regarded as a new super machine."* Not a bad definition of network computing. No wonder we have problems managing these monster webs of "super machines."

What would a dream machine look like? Exhibit 4.3 lists the functional characteristics of a network computing dream environment. First, it would be consistent and predictable, delivering fast, reliable service to everyone in the enterprise, whether on the road, at home, or in the office. Standard network services would include universal e-mail for everyone; an intranet for electronic publishing and innovative enterprise applications; and a standard platform for delivering network computing to every desktop and workbench. The dream machine would be flexible and easy to upgrade as new technology came on the scene. A central Help Desk would manage network utilities and answer every question with clear, accurate instructions. Access to the outside world would be seamless, efficient, and universal with protection against intruders. Finally, the dream machine would make electronic commerce as easy to use as the telephone and it would support transferring large files to anyone in the world.

A digital organization allows you to move ones and zeros from anyplace to anywhere at anytime. A dream machine provides the vehicle. Can we realistically and economically build a dream machine? Yes, the

---

* Alvin Toffler, *Future Shock* (1970), New York: Random House.

Exhibit 4.3
The Dream Machine

Consistent Architecture and Configurations
  Multiprotocol network using IP addressing.
  Intranet, universal e-mail and groupware.
  Intelligent premise equipment.
  Broad bandwidth premise wiring with efficient site traffic.
  Software restricted to servers.

Central Help Desk
  Staffed with highly trained technicians.
  Integrated problem and change management with inventory.
  Diagnostic and network management tools.
  Problem resolution tools.
  Centralized electronic software distribution.
  Centralized server backup.
  Electronic inventory.

Internet and VAN Access
  Web home pages, both external and internal.
  Double firewall.
  Virus protection.

Utilities and Features
  Single logon facility.
  Enterprise directory.
  Sophisticated electronic commerce capabilities.
  Bulk data transfer capability.

Applications Make the Dream Machine Worth Owning

---

technology exists, it's only a matter of will. The installed base of equipment and related processes, combined with an ingrained culture, presents the biggest hurdle. Before reviewing AlliedSignal's strategy for building a network computing machine, let me whet your appetite by further explaining the target.

Multiculturalism in the Network

You cannot build a homogeneous computing environment, and you don't want to. First, the cost would exceed the national budget, but more important, you cannot guess where technology will go. Since you need the freedom to adopt anything attractive that comes along, you want a heterogeneous network that allows you to mix and match your installed base with the new technology. Besides, different platforms provide advantages for different types of computing. So the first criterion of a robust enterprise network is the ability to handle multiple protocols, support disparate devices, and communicate with whatever your customers and suppliers choose to adopt. Luckily, the macrotrend toward Internet Protocol (IP) addressing and multilingual agents makes this feasible.

An IP addressing scheme roughly correlates to your street address with a zip code, street number, and street name. Agents are small pieces of software that reside on network equipment to monitor and report activity to a central network management server.

A digital organization has universal, comprehensive and consistent e-mail, groupware, and Web applications for the entire enterprise. These capabilities, combined with a modern voice system and videoconferencing, provision the organization with the breadth of communications necessary to leverage the entire organization's skills. However, you still must communicate with customers, suppliers, governments, and academia, and they may not pick your specific suite of products. Internet/Web technology may soon make this problem go away, but for the time being, your enterprise network needs an interpreter between your systems and the rest of the world. So provision your network with one of the many products available to interpret discourse between yourself and those outside your bastion of enlightenment.

Every site in a digital organization has free-flowing traffic on the LAN segments with convenient access to I-Way on-ramps for extending beyond the company's internal network. Everyone can easily connect to the network when they venture away from the office and things stay in sync between their desktop equipment and their laptop device. Every piece of equipment is able to communicate back to a central Help Desk.

Premise wiring schemes provide abundant bandwidth and make moving people as easy as switching vacuum cleaner plugs between rooms. All software resides on LAN servers so updates can be electronically distributed and no one violates license agreements.

Making this all happen requires that the architecture, hardware configurations, software, network tools, and addressing all conform to enterprise standards. It may seem like a contradiction to insist on strict standards while simultaneously supporting a heterogeneous environment—it is not. Standards limit diversity to reasonable and justifiable levels and put in place the necessary interfaces to make different equipment work together. Without standards, chaos reigns and people all do their own thing to such an extent you will never leverage your entire organization with digital technology.

Kit and Caboodle

The Help Desk must be an integral part of an infrastructure upgrade initiative. The ideal Help Desk should be provisioned with the latest tools and staffed with competent, highly trained, and completely idle people. Since our endless endeavor to incorporate ever more sophisticated systems makes this impossible, converting the Help Desk from a reactive to a proactive organization presents the next best option.

You can't just apply network management tools and utilities and expect improvement in your network performance. These contemporary tools require the site infrastructure to be compliant with their protocols. That's why you need a comprehensive plan to do both simultaneously. When a site, segment, or device is noncompliant, the Help Desk goes blind and the utilities become inert. If you want a digital organization, you need to do it all.

An integrated problem/change management system provides the most valuable assistance for Help Desks. In today's world, this must be an intranet application. Mainframe-based problem systems provide the same functionality, but personal computer technicians dislike using mainframe systems. If you still have a mainframe problem, change management system, switch it out for one that everyone finds acceptable.

Network problems lend themselves to pareto analysis, so make sure *every* problem and change gets logged and analyzed for trends.

The problem/change system should include inventory so when a call comes in, personal identification, location, configuration, and past problems pop up automatically. It also helps to integrate an expert system into the process, along with procedure and process descriptions. Essentially, you want every piece of information possible at the fingertips of a highly qualified technician so problems don't need to be passed up to second level support. Don't make the mistake of putting increasingly qualified people at successively higher levels of support, and keep your best technicians in the field. Your best people should be on the Help Desk. You want first-call resolution, and using experts to answer the phone provides the only way of achieving this goal.

Change management must include every hardware, software, and configuration change, because when a LAN goes belly-up on Monday morning in Indiana, your Help Desk in Oklahoma needs to know an upgrade occurred on Sunday for a field sales force application running on that LAN.

Network management tools allow a Help Desk to see a problem and react before the user places a phone call. No single product does the entire job, so you need to pick a suite that will undoubtedly have some redundancies. Since each product wants a separate console, console automation becomes mandatory. Automate repetitive tasks, integrate alarms, integrate Help Desk applications, and consolidate all these functions to a single GUI console. Your Help Desk analysts will thank you.

We need a better name for the Help Desk. These nerve and control centers have assumed responsibility for centralized network services, including electronic software distribution, mail administration, LAN backups, change management, inventory, directories services, problem resolution, and sundry other tasks. If you haven't reviewed your compensation and job grades for these positions recently, then take a close look before your Help Desk becomes a recruiting source for your competitors.

Networking the Neighborhood

As you build your digital world, you need to connect to the outside so your intranet truly becomes part of the global Internet. You want to do this carefully because there are hooligans out here eager to cause mischief. Most just want to play with your toys, but a few like to inflict real damage. Firewalls provide the same protection as locking your car—they stop amateurs and discourage professionals. The best you can aim for is making your network more difficult to break into than your neighbors. True professional hackers can break through any security system until your firewall provider or in-house technicians catch on and close that particular door. It's a never-ending race, so be vigilant and keep your eyes open for intruders.

Connecting to the outside world also increases your risk of catching a virus. Virus protection only works on the latest infection, not new mutations, so stay on guard. Restricting external access to a central control organization gives you as much protection as possible from both burglars and their sneaky counterparts that use virus proxies to intrude. If you have business units establishing their own Internet connections, shut them down as fast as possible.

Lost productivity presents another Internet risk. Surfing can be a big time waster, and worse, your employees may venture into the bad part of town. Another firewall function is blocking access to obscene material on the Internet, but don't restrict only sexual content, stop access to everything that has no business purpose. I once had an employee protest because we blocked the entertainment portion of the Internet. I couldn't believe someone was foolish enough to complain about no longer being able to follow the Soaps.

You also should check to see whether employees use the company credit card to buy Internet access from home. AlliedSignal found hundreds using this end-around and put a quick stop to it. People knew about the poor security on the Internet, so they decided the company should bear the risk of theft. They used the company's credit card, rather than their own, to travel along routes frequented by highwaymen. A new policy stated that home access to the Internet for business purposes could only be done by dialing into the CTC's facility. AlliedSignal also notified the access providers directly to cancel subscriptions.

Web pages and browsers did the same for the Internet as automatic transmissions did for the automobile—made tooling down the road easy for people with two left feet. Usually, companies find the internal use of this technology more valuable than cruising the interstate. Today, most companies establish their own intranets to distribute information. Web technology is the first new publishing media in five hundred years and it may eventually exceed the importance of the printing press. Web and hypertext technology produce documents, with an interactive capability to obtain more detailed and imbedded multimedia that the reader can engage or ignore.

Some types of publications, like encyclopedias, will seldom see the printed page again, but despite the hype, Internet/Web publishing will never replace the novel or anything else designed to be read in sequence. Instead, whole new categories of publishing will emerge to take advantage of the ad hoc nature of the media. In the business arena, many documents lend themselves to Web pages. AlliedSignal put the Controller's Manual, phone directories, executive briefings, newsletters, Security Manual, organization charts, site location maps, and lots of other things on Web pages. Revisions and updates get immediate attention, the media holds interest, people can comment directly to the authors, only those interested pull down material, and the cost for publication and distribution decreases dramatically.

Similarly, Web applications do not replace systems designed to discipline a process. Again, the ad hoc nature does not direct the user in a particular sequence. The most interesting business applications allow people to do business with you remotely, either from home, the road, or another business. Chapter 8 gives some hints on the potential use of this technology to save money and increase sales.

A Fully Equipped Network

Addressing presents the biggest problem with LAN-based e-mail. This mundane administrative function makes the difference between a cranky e-mail system and one that functions as smoothly as the telephone. You need two things to support an enterprise directory: a global directory software capability and sufficient staff to keep it current and accurate. New e-mail software includes this feature and if you're still on an old release, it's time to make the upgrade. Don't slight this critical function by not allocating enough people to do the job right. After all, you don't want people surfing your intranet to find someone or something. It's bad enough when they venture out onto the Internet.

Another underrated feature of a good intranet is a single logon facility. Don't irritate people by making them identify themselves every time they switch computers. The purist like to say the network is the computer. Be sure to purchase one of the software utilities that actually makes the network look like a single computer.

Sometimes people, especially engineers, need to send very large files through the network. You need your own cyber "parcel post" to keep these watermelons from clogging your network pipe. E-mail doesn't do a reliable job of delivering bulky packages. Special utilities exist to meet this specific need and every well-equipped network needs one. Another suggestion: As soon as someone develops a good network scheduling package, buy it. Network technology desperately needs a utility to schedule these bulk deliveries late at night while people are snuggled deep in their duvets.

Increasingly, business transaction documentation will be done electronically—you will be able to do business with your customers and suppliers without paper. To make this a reality, build an EDI processing center as part of your network operations. Centralizing this service saves money and reduces setup time.

Building the Dream Machine

We've now defined the machine of our dreams, but how do you go about actually building one? AlliedSignal initiated a major program to build a dream machine similar to the previous description. We named the initiative GlobalLink. An early briefing defined the goal as, "A reliable global I/S infrastructure that connects every AlliedSignal desktop and workbench with technology that makes applications easy to use and less expensive."

We worked hard on this definition. It includes all AlliedSignal world-wide sites, does not exclude the factory, and sets objectives of connectivity, ease-of-use, cost control, and reliability. Exhibit 4.4 shows the steps of strategy, but it could really be described in a single word—simplify! Making network computing function on an enterprise basis requires taking the complexity out of the infrastructure. Simplification requires central control, so AlliedSignal's first step involved assigning responsibility to the CTC for designing the architecture, setting the standards, and executing the actual upgrade of the infrastructure.

Architectural design started with defining all the elements between the desktop and the most far-flung device in the network. Exhibit 4.5 shows the CTC's definition of the first and second hierarchical layers of network components. The third layer, not shown, included over a hundred elements. Once we knew how many Lego® blocks we needed, we sent out a survey to find out what was already installed. The surveys confirmed that we had not overestimated the scope of the problem. If

---

Exhibit 4.4
AlliedSignal's Infrastructure Strategy

1. Centralize management responsibility.
2. Classification of infrastructure elements.
3. Determine installed base.
4. Define architecture and establish standards.
5. Perform gap analysis and maturity model.
6. Determine cost of required upgrades.
7. Upgrade infrastructure.
8. Implement process improvements with upgrade.
9. Enforce standards.
10. Implement and evergreen policy.

Significant Resistance Due to Major Culture Shift

Exhibit 4.5
AlliedSignal's Network Components

I. Desktop:
    Personal computers.
    Dumb terminals.
    Engineering workstations.
    Host emulation packages.
    Desktop operating system.
    Custom applications.
    Agents.
    Office suite.

II. Site Infrastructure:
    Wiring/cable.
    Hubs/concentrators.
    Switches/bridges.
    LAN segmentation.

III. Mail and Groupware:
    E-mail.
    Groupware/bulletin board.
    Internet/intranet.

IV. Servers and Networks:
    Server/network operating system.
    Server hardware.
    Software residence.

V. Management Tools:
    Addressing protocol.
    LAN backup.
    File transfer.
    Security.
    Virus protection.
    Software distribution.
    Network management.

anyone ever sold it, we bought it and found some way to wedge it into our network.

Another aspect of this analysis included the definition of a maturity model and the classification of every company site as A, B, or C. An "A" site didn't need our help because they already complied with the architecture and standards. "C" sites hired Rube Goldberg to design their monstrosities and Mickey Mouse to do the installation. "B" sites fell somewhere in the middle. Fortune didn't smile on us; we had no "A" sites. Oh well, we had anticipated that going in—otherwise we wouldn't have needed a corporate initiative to clean up the rubble.

Once we knew what we had versus what we wanted in our dream machine, the next step involved a financial analysis to determine the cost of switching out all the noncompliant equipment and software. The big-dollars warning provided by this analysis caused another design iteration to shave, trim, and nip cost. Then we got together with our procurement people, telling them we needed a few million more cut through their buying prowess. Surprisingly, this worked rather well as they leveraged AlliedSignal's size and standardization to obtain steep discounts from our previous corporate pricing.

Execution remains the test of fire. Independent of your strategy, the problems encountered in execution remain similar. Next, I'll explain how AlliedSignal executed these initiatives while retaining our sanity, sense of humor, and vocation. First, I want to point out our biggest disappointment. We underestimated the mess on the desktop. We originally thought we would blow into town with a high-powered team, perform a "wellness assessment," rewire where necessary, realign the LAN segments, add server capacity, switch out obsolete equipment, clean up addressing and other housekeeping duties, and quickly move on to the next site. We discovered desktops so convoluted that we had to make a deskside visit to every workstation to straighten out the mishmash of software, files, and formats.

We didn't have enough people to make these visits and keep on schedule. So we had to head back to the senior leadership for more money to contract the additional technicians needed to keep the program on an 18-month track. Out of all our technology initiatives, this was the only time we had to go back and beg for additional funding, and it was not a pleasant experience. Thankfully, we were under-running our capital budget, which helped offset part of the increased expense spending.

More money didn't solve all our problems. Unlike our other technology initiatives, GlobalLink ran up against some severe cultural resistance. When you close down a data center, you usually upset only the operations manager and the site director for information technology. Start making changes at workstations and you can raise everyone's ire.

Exhibit 4.6 shows the barriers identified at the start of AlliedSignal's GlobalLink program. Site resistance to corporate interference with internal affairs represented our most difficult hurdle. Pulling this project off took continued top management involvement, as opposed to other initiatives that received relatively little attention after approval of the business case.

The secret to keeping momentum is flawless execution. I need to make a final point in this chapter—you don't want to do this twice. Since the technology moves fast, you need to constantly reevaluate standards, then strictly enforce them, and set billing rates and depreciation schedules to reflect the true cost of network computing. An evergreen philosophy will drive you to bill out services at a rate that can fund continual upgrade of the infrastructure. People will yell and scream because they never recognized the real cost before, but survive this one temper tantrum and you can keep your network infrastructure contemporary.

Exhibit 4.6
Barriers

Installed base.

Low level of support, both central and site.

Decentralized management responsibility.

Site resistance due to loss of control.

High performers leaving.

No published architecture.

Building a digital organization requires diligence, good management, and organizational leadership, but we're all facing the same challenge. I returned from a conference one time and told my people I had good news and bad news. The good news was, "Everyone has the same problems that we are experiencing." The bad news was "If you don't like the situation here, you can't escape it by going someplace else."

CHAPTER 5

TOUR DE FORCE

Manage Your Execution

Well done is better than well said.

—BENJAMIN FRANKLIN

"Where are they?" Bossidy asked.

I quickly responded that they were in the building we had just left.

Bossidy whirled around and practically ran out of the data center. This was back in the early days, when we were still learning how to execute technology initiatives. Our second data center consolidation had just been completed on schedule, and Bossidy had come to Arizona to see the new Computing Technology Center. We had done a quick tour of the office area and then walked the short distance to the center. On entering, I mentioned that the consolidation team was conducting a postmortem on the past weekend's activities. This comment sparked his question and quick exit. Damn. I had been looking forward to showing off our new facility, especially the control center that looked like a set from *Star Trek*.

But Bossidy showed no interest in a bunch of hardware; he wanted to talk to the people who made it happen. I sprinted to catch up as he began rapid-firing questions:

"What's the purpose of the meeting and who'll be there?"
"What kind of problems did you encounter?"
"What do your customers think about the operation?"
"Is there anything you'd like me to say?"

I told him the meeting included about 40 to 50 people, all of whom had some part in closing the latest data center and moving it to Tempe. In the beginning, we had established a standard practice of conducting a postmortem after every major event. The meeting agenda included reviewing open issues, identifying what went wrong, and defining corrective actions. The meeting would document our "lessons learned." The customer impact would be reviewed and methods defined to reduce any disruptions to business operations. I asked him to compliment the people on a job well-done and recognize all the extra hours they had invested to ensure a smooth transition.

Bossidy barreled into the standing-room-only meeting, introduced himself and asked, "What is the most important lesson learned from this consolidation?"

I held my breath, wondering if anyone would have the nerve to answer an unexpected question from the chairman of the Board. I didn't need to worry. Someone immediately responded, "We need to remember what we did right, as well as look at what went wrong. This time we made different mistakes, and many of them we did right the first time. So we need to nail down our good processes, as well as correct our errors."

This answer represented exactly the kind of thinking Bossidy wanted to hear: *Hone your processes and make them smoother, faster, cheaper.* Continuous improvement means working smart, not thrashing about, hoping frenzied activity will make success happen.

The rest of the meeting proceeded in an informal, chatty fashion with Bossidy showing keen interest in every aspect of the operation. After the exchange became comfortable, Bossidy explained how the data center consolidation program fit with the company's overall business strategy. He shared his vision for AlliedSignal and answered a few questions about the direction of the company. Everyone in the room had heard these words many times before, but Bossidy never passed up an opportunity to reinforce his message. Why? Because he understands the first imperative of execution— keep the entire organization focused on clear objectives.

The Eight Imperatives of Execution

Execution is the nemesis of lofty plans. Too many good strategies languish because leaders underestimate the difficulty of altering an organization's behavior. You need to go beyond the conceptual planning stage, using every tool at your disposal to drive your plans to fruition. Strategies must be sold to the masses, enthusiasm constantly rekindled, and everyone's attention riveted to achieving the objective. The arduous task of merely keeping pace makes it too easy to get mired in the status quo. Eight imperatives for solid execution are:

The Eight Imperatives of Execution

1. Focus the entire organization on clear objectives.
2. Use management principles consistently.
3. Continuously improve execution.
4. Focus on customers.
5. Move with speed.
6. Benchmark operations.
7. Manage supplier relationships.
8. Overcome resistance.

Clear Objectives

Corporate initiatives need a clear set of goals and objectives. Use the military example for setting objectives: Take that hill, using this many troops, and get it done within a specified time. Then lead with behaviors you want to see throughout the organization. Establish aggressive schedules and focus the organization so everyone pulls in the same direction. Stay consistent, because shifting priorities destroy momentum. *Constantly communicate the mission, schedule, budget constraints, progress, next steps, and never waver or vacillate.*

Corporate technology initiatives tend to be political, fast, technically challenging, and stressful. To focus an organization requires more than an inspirational kickoff speech; you need to use every managerial lever to keep on track. Many of these will be discussed in the next section, but here I want to emphasize the need to constantly communicate a consistent message. You may believe everyone knows the strategy, but bringing it up at every opportunity reminds people the agenda still has priority. Thank people for making progress, ask questions about near-term milestones, communicate what other parts of the organization are doing, discuss any revisions in the plan and find other relevant ways to let people know this still has your highest concern.

Use Management Principles Consistently

The opportunity to recruit an entire staff from scratch does not happen often in the business world. Most of the time, you need to motivate an existing staff, weed out a few poor players, and selectively replace them with new talent. Frequently you hear that motivating people requires leadership qualities. True, but charisma is not a prerequisite. I've worked for successful leaders that didn't exude charm, but they knew how to use standard management principles to move an organization in a desired direction.

Managers possess a large arsenal of techniques to alter an organization's work patterns, but using them all consistently magnifies their effect. The most important tools are pay, performance reviews, rewards and recognition, and celebrating success (Exhibit 5.1). We all know we

---

Exhibit 5.1
Motivate the Organization

Provide pay-for-performance.
Use effective performance reviews.
Offer reward and recognition.
Celebrate success.

---

should pay attention to these basics, but we frequently get distracted by alligators nipping at our heels.

Money Talks

Money is important, but its negative impact can dwarf the positive influence. When people receive a raise, they soon accept it as their due. However, compensation that is perceived as unfair, ruins the spirit of the organization. Managers swamped with other commitments devise merit plans that barely differentiate the best and worst performers. Many managers distribute raises evenly to avoid confrontation with poor performers. When you enforce a pay-for-performance philosophy, it will automatically improve performance reviews because managers must explain to certain individuals why they didn't receive the increase they expected. Building motivation to execute tough technology initiatives becomes easier when people earn their raises and managers give honest performance reviews.

Promotions and terminations need to pass the swivel-neck test. When you terminate someone, you want everyone to quickly turn and say, "It's about time!" Conversely, when a promotion is announced, everyone should recognize that merit accounted for the elevation in rank. They should quickly turn and say, "It's about time!" Try to visualize the organization's response in advance of taking action; you'll be surprised how much it helps with both the action and the timing.

Reward and Recognition

In addition to salary and benefits, modern organizations effectively use rewards and recognition to motivate behavior. Larry Bossidy wants exemplary performance by every employee, so upon arriving at AlliedSignal, he instituted a reward policy for nonincentive employees that can go as high as 10 percent of annual salary. Each operating unit has a great deal of flexibility in designing their own reward and recognition program. The CTC designed a three-tier system that includes a management recognition program, peer-to-peer recognition, and rewards issued by management.

The management recognition program has three levels: a free lunch in the cafeteria, dinner for two, and a leather notebook embossed with the CTC and AlliedSignal logos. The lunch certificates are freely distributed as quick recognition for doing something beyond the daily grind. It's a nice way for management to say, "Thanks, I appreciate the way you handled that situation." There are two ways of earning a dinner certificate: receiving a customer letter of appreciation or putting in extra hours to meet a commitment. A dinner for two also thanks the other person who may have been inconvenienced by the extra hours. If someone comes up with a really innovative idea, resolves a major customer problem, or invents a superior business process, they receive a leather notebook with a handwritten commendation letter on the first page, a copy of which goes into the recipient's personnel file. People proudly bring these executive-style folders to meetings, encouraging others to leap tall buildings to earn a notebook.

One day, a small group came to see me to complain that rewards and recognition all came from the top. When I asked what they would suggest, they asked permission to design a peer-to-peer program. I liked the idea and told them go ahead. This self-formed team came back with an excellent recognition program that included a budget, procedure, and an automated voting system using e-mail. Each CTC employee receives three votes a quarter that they can award to anyone they feel deserves recognition. The happy recipient finds a thank-you in their e-mail along with an electronic 10-dollar bill. Each quarter, we publicly recognize the people and teams with the most votes and at the end of the year announce the top winner at the Christmas party. Besides the monetary award, the person with the most votes during the year receives other perks along with the prestige of being recognized by his or her peers. This great program is extremely popular, inexpensive, and a great motivator.

The third tier represents the real money. Using the Bossidy-approved program, the CTC awards cash to the people who made a major success happen. Whenever we accomplish a difficult commitment, a management team assesses the contributors and decides on award amounts. The rewards are publicly distributed in meetings of the entire CTC staff, with the amounts held confidential by a sealed envelope. People soon learn that meeting our challenging assignments does not go unnoticed, nor unrewarded.

Party On

"Celebrating success" is a very effective way to reinforce the strategic direction. Use celebrations selectively to recognize significant milestones or measurable progress toward a defined set of goals. Keep the event focused on the accomplishment. Don't give any speeches on the next agenda item; just relax, have fun, and celebrate success without worrying the next hurdle.

I'm not a party kind of guy, so I searched the organization for someone who enjoyed planning celebrations. One of our best parties had a *M°A°S°H* theme. On the day of the event, everyone received two olive-drab T-shirts, one for the employee and one for a guest. The shirt had the M°A°S°H logo, with the words "Made Another Success Happen" printed underneath. We rented a local air museum hangar with World War II planes and memorabilia, accented it with USO decorations, and hired an Andrew Sisters knock-off group for entertainment. Everyone had a great time.

New Recruits

You always have at least a few opportunities to refresh your organization. Turnovers, work-outs, and retirements always provide a spattering of openings, but corporate technology initiatives frequently allow larger recruitment efforts because of the centralization of a dispersed staff. Exhibit 5.2 lists the recruitment steps I found effective.

Start with good job descriptions. You advertise when you sell something, so approach the job description as advertising, not as a bureaucratic chore. Don't use too much Madison Avenue, or you'll build false expectations. I'm specific on the function and responsibilities, and include a description of the organization's culture. The job description, or posting, presents an underestimated opportunity to mold behavior. State exactly what you expect, so people who don't fit can eliminate themselves.

Exhibit 5.2
Recruiting for Technology Initiatives

Sell the initiative, organization, and culture.

Use job descriptions and postings as advertisements.

Immediately broadcast new recruits.

Use a three tier nomination process.

Partner with human resources.

You need a nomination process to make sure the good catches bob to the surface. AlliedSignal uses a three-tiered approach: business nominations, functional nominations, and self-nominations. By chumming with good job postings and trailing three lines in the water, we snag everyone not hunkered to their desk so closely they'll probably never rise to the surface. Once you make a good catch, announce your trophy far and wide to draw other good candidates to the program. Competent people have a reputation within their discipline, and they attract soul mates into the selection process.

Technology initiatives in large organizations require more than recruiting a bunch of sharpshooters. A good staff makes the job easier, but selecting them properly increases your odds enormously. First, recruit people from as many of the target sites as possible. People from the various business units know the operations, personalities, and politics of their previous home. This becomes invaluable when you start planning. Business units feel better knowing they have a few of their own people on the ground to protect their interests. We also make it a practice to hire at least a couple people from each business acquisition.

If, during the recruiting process, it seems impossible to hire anyone from a particular site—watch out. All of our initiatives went well, but a few had some ugly moments. In these instances, no one had accepted a position in the CTC. The site executives painted such a negative picture of the initiative that a resistant attitude permeated the organization, persisting through operational takeover. If you do a good job recruiting and nobody steps forward, take this as a signal that you have deep-seated resistance within the business.

I have learned I need Human Resources on my side. When I scoured the company intent on attracting the best talent, I soon discover that these people carried pay and titles reflecting their star quality. Human Resource professionals design policies to preclude building top heavy, highly compensated organizations. When you build a small central staff to support enterprise technology initiatives, you need a higher concentration of talent. I didn't want to just recruit these people, I wanted to keep them motivated. Sometimes you must work with your Human Resource people so they understand you need a little leeway on compensation guidelines.

A common scenario includes building a central staff to support the progressive rollout of a corporate technology initiative. The recruited people support their current business with a similar, but decentralized service. As you recruit these people, immediately relocate them to their new home. You'll hear the arguments that site support will crash and burn if you pull these people out, but don't listen. When people have to move to retain their jobs, they frequently accept, but start looking for new employment in the old neighborhood. When it's time to go get them, they're gone. Besides, you want to start building your culture and you need to corral people to do this effectively. If support issues arise prior to taking over a site, dispatch people from your central staff.

One last point, you want a slightly different skill set for corporate technology initiatives than for divisional solutions. Enterprise technology initiatives tend to be centrally directed and you need people who can communicate over the phone or in front of a group. Besides technical knowledge, they need good interpersonal skills. The combination may be hard to find, but the additional searching pays off.

Continuously Improve Execution

Enterprise technology initiatives need good execution plans. However, there needs to be a balance between trusting in serendipity versus planning until the Cossacks have breached the parameter. We brought a consultant in once to see if his firm had a better way to plan large initiatives.

He spoke eloquently about his references, comprehensive planning methodology, tracking systems, and ability to drill down to the smallest detail with his customized software. We asked him how long it took to plan a large corporate initiative. He said six months, and we all burst out laughing.

"What's so funny?" he asked.

We told him we didn't take six months to plan anything. "How in the world could you take that long," I asked, "You must have planned the entire initiative to completion."

"Of course, how else would you do it?"

We explained that we only planned the next hurdle. We spent little energy worrying about what stood behind the obstacle immediately in our path. He protested that we could make serious mistakes by not looking further ahead. In our experience, extensive planning didn't make sense in the hyperactive computer field and the few errors we did encounter were easily reversed or overcome.

Our philosophy was to get momentum, and then keep it by driving to completion the next thing in line. Marathon runners don't think about the distance to the finish line; they concentrate on the next few steps. The CTC does extensive planning, testing, and rehearsing for the next major milestone, but we don't worry about the inchstones that lie beyond the horizon of our next assignment. If you miss a milestone, it cascades throughout the initiative, destroying schedules and budgets.

Concentrate on developing methodologies that the organization can learn and feel confident about repeating when the time comes. Changing technology, shifting priorities, and corporate restructuring disrupt long-range plans. You can't anticipate everything, so concentrate on making the next step go smoothly.

The data center program included eleven consolidations of existing centers into the CTC. Besides being technically challenging, these moves required horrendous logistics. The lessons learned from AlliedSignal's first corporate technology initiative served the CTC well in the follow-on programs. I attended every center consolidation, even though they occurred on weekends in the dead of night. During the first consolidation, a small army tried to fit into the control center. The place buzzed with activity, and it looked like a casting call for the Keystone Kops. We muscled our way through, but there had to be a better way.

By the fifth consolidation, a novice observer wouldn't have noticed anything out of the ordinary. Two or three people would be keying commands into the system. If they talked at all, the tone was quiet and unhurried. At some point, two or three new people would enter and casually look over the shoulders of the people working. In a few minutes, the first group would rise, stretch, bid the newcomers welcome, and give them a quick status. The second group would take the chairs and start their own set of commands as the previous occupants left, their work complete. It only *looked* quiet. In reality, tens of millions of computer instructions and gigabytes of data transferred across the nation.

Even the transfer of the physical tapes looked like a nonevent. The racks in the old center were shrink-wrapped without removing the tapes and a simple razor blade cut away the plastic when they arrived in their new home. Batch production processing started as soon as the 16-wheeler pulled up outside the CTC. Everything moved so smoothly that our customers didn't even know what happened. After one of the later consolidations, a manufacturing vice president stopped the site I/S director to ask when the data center consolidation was finally going to happen. To his surprise, he discovered it had occurred two weeks previously. Professionals make the difficult look simple because they practice until it seems effortless.

The postmortems provided the mechanism for making each successive execution better than the preceding one. We used checklists to make sure nothing slipped through the cracks. Test plans and procedures isolated problems before they impacted the big event. Automated procedures allowed analysts to initiate a single command, starting a whole series of actions. Practice made the logistics function faultlessly and contingency plans kept the unexpected from disrupting a smooth execution.

The Best Laid Plans

A couple of incidents drove home the importance of contingency planning. Whenever we did a consolidation, two sets of data restore tapes were flown in on separate aircraft, a primary set and a backup. Since we seldom went to the backup tapes, we felt pretty confident—until we consolidated our center in South Bend. During this move, both the primary and the backup failed for the same volume of data.

Not even a ripple of panic infected the control room. We always left two people at the site being consolidated until production resumed; one remained at the center, while the other caught some sleep. People immediately went into action. Someone pulled up the airline schedules discovering that the last flight out of South Bend took off in less than 30 minutes. Another person called the center and told the analyst which volume to copy, instructing him to get it done quick and take it out to the curb. A third person called the hotel, woke the other analyst and told her to drive by the center, pick up the tape as she drove by, and make haste to the airport. Despite this practical demonstration of Murphy's law, South Bend production started up less than two hours behind schedule. I did give the analyst a hard time for not duct-taping her precious cargo to the aircraft's black box so we could find it in the event of a crash.

We avoided another major gaffe by watching weather reports. Living in the Valley of the Sun, locals assume every day will be just like the day before. Luckily, the project leader picked up a storm alert and rescheduled a consolidation to start three hours earlier than planned. Our plane took off a mere 20 minutes before the entire East Coast shut down.

The obsession with continuous improvement works for data center consolidation, deploying new networks across the globe, standardizing the desktop, upgrading infrastructure, or any other corporate technology initiative. These all require going out and doing something at one site after another that can disrupt business. Each event should have two objectives; smoothly implementing the change, and executing it better this time than any previous time. Accomplishing this requires rigorous reviews of operational activities to scrub history for lessons that will make the future easier than the past.

Focus on Customers

Managing a computer service organization means having an unlimited number of bosses. Everyone feels free to demand exclusive service. Customer difficulties arise because you first take something away from them, and then force them to change to a new corporate systems standard not of their choosing. In return, they feel justified in demanding improved service with rapidly declining prices. This makes the job more difficult, not impossible. If you keep a customer focus, old resentments eventually fade and you win them over.

Customer focus must be approached differently for internal customers. First, it's not a true customer relationship. They normally can't buy from anyone else, and you can't hang a sign that reads, "We reserve the right to refuse service to anyone." To me, internal customer focus means empathy for their problems and working with them to make those problems go away.

Customer focus is more than an attitude. The design of an organization must meet the customer's needs. A best-in-class help desk provides the first line of defense. You need to equip it with a good problem/change management system and invest in extensive training. The Help Desk Institute says a superior Help Desk can make a mediocre operation look good and vice versa. Pay attention to this critical function and elevate its status so technicians consider it a promotion to move to the Help Desk.

When the CTC was formed, I invested the most care in the design and staffing of the quality and customer service organization. I believe it is crucial to arm customer service with the resources to properly resolve customer problems. Training everyone to be customer-oriented helps, but you also need to build an effective customer service organization with the breadth of authority necessary to address customer needs. Initially, the cost of this group caused some second thoughts, but after few years' experience, it has proved to be a smart move.

Within the CTC, the most critical element of the customer service organization was the establishment of a small cadre of client executives. Client executives have senior positions and represent an assigned set of business units. Their primary role is satisfying the customers' need for personal service. They visit the sites, hold weekly voice conferences, and provide prompt responses to their customers' requests for service. When the technical managers complain that the client executives take the customers side too strongly, I know they're doing their job.

Improving financial performance represents the most pressing problem of the CTC's customers. The accounting function reports to customer service so that the client executives can respond authoritatively to customer billing issues. As listed in Exhibit 5.3, other responsibilities for this important function include service level agreements, metrics, communications, strategic planning, budgets, contingency planning, billing, and instilling a customer attitude throughout the entire organization.

Instilling a customer-oriented attitude represents the most important function of the customer service group. Holding a function permanently responsible for maintaining a customer culture works better than putting the entire organization through one-time training to heighten customer sensitivity.

Move with Speed

Budgets for technology initiatives assume tough schedules; one misstep not only puts the program behind schedule, but destroys the economics. Speed forces innovative approaches and solutions, drives decision

---

Exhibit 5.3
Customer Service Functions

Service level agreements.

Metrics and benchmarking.

Communications.

Strategic planning.

Budgeting.

Contingency planning.

Billing.

Instilling a customer attitude in the organization.

---

making, and overcomes resistance. Besides, if something is worth doing, then the sooner it gets done, the better.

AlliedSignal demands a comprehensive business case for every initiative. The faster the implementation, then the quicker the benefits accrue and the easier it is to gain approval. AlliedSignal also liquidates all costs against the businesses. Since most initiatives have start-up costs, a quick pace reduces the problem of incurring expenses prior to delivering a billable service. (Matching cost with billable service presents the toughest planning job. Ramp-up too soon and you jeopardize the program's economics, cut it too fine and the risk becomes unacceptable.) Acceptable payback periods in corporate America are becoming ludicrous, but neither you nor I can change that. Accelerating schedules provides the only way I know to meet these stringent requirements.

Decision Time

Very tight schedules force fast decision making. Look at some of your own initiatives. I'll bet that getting decisions on key issues represents the most significant contributor to elapsed time. Computer people are analytical. Their comfort level requires extensive analysis prior to making a decision. Technology moves before analysts finish testing and projects end up in the recycle bin before they get out of the lab. Despite any discomfort, you need to force rapid decisions to keep pace. Apply Pareto's law—get an 80 percent solution today, rather than a 100 percent solution never.

Imposing time limits presents the best way to force decisiveness. (It works for game shows and it will work for you.) What are the penalties? A few bad decisions. One time we sat in a conference room and listed 11 bad decisions during the early days of the center—3 lost-cost bidder errors, 4 technical mistakes, and 4 product selections that didn't make the grade. Given the thousands of decisions it took to get us started, we felt pretty good about the count. We survived these mistakes because someone noticed them early on and either reversed them or developed an effective workaround.

If you want to move fast, you need a culture that rewards recognizing bad decisions and taking quick corrective action. Decisions drag out because people fear accountability. Change the fear button; put it square on the forehead of anyone who tries to disguise a poor decision and reward people for bringing forward decisions that need to be reviewed. Speed doesn't kill, crashes kill. *Avoid collisions by constantly reevaluating decisions and changing course when necessary.*

Mickey Mouse to the Rescue

Americans have an innovative spirit. Historians tell us it reflects heritage from the frontier settlers who had to make do with what they had readily at hand. Speed takes advantage of this spirit by forcing innovation and resourcefulness.

During construction of the data center, the facilities director came to see me in an agitated state. It seems he had a problem with his boss, the vice president of manufacturing. The building selected for the new center—a large, two-story multipurpose facility—had an office area at the front and manufacturing in the rear. The layout included a receiving dock at the back of the building, with our raised-floor equipment room located on the second floor in front. Our construction activity disrupted production receiving, clogged the manufacturing aisles, and monopolized the freight elevator.

"I've been ordered to restrict your construction activity to third shift and Sundays," the director said. "I never anticipated how much material and equipment you guys needed. The chaos in the manufacturing area is causing us to miss shipments."

"How about the construction schedule?" I asked.

"This will delay things, but I haven't figured out how much yet."

I got up, preparing to leave my office, saying, "Sorry, no schedule slips, you'll just have to figure something out."

I left immediately, leaving no room for discussion. I felt sorry for him because he was between his boss and a corporate function, but I couldn't extend his schedule and still make my commitments.

The next day, my operations manager called and said I should walk over to the center and see what facilities had done. Oh, oh, I thought, what now? As I approached the building, my jaw dropped—a huge gaping hole now adorned the front of the second floor. Facilities had punched through the concrete and positioned a crane to lift material directly into the equipment room. I loved it.

Planning on the Fly

A Friday meeting with a U.K. supplier had just started when my European director received an emergency call on his cell phone. He nervously pulled me into the hall to tell me that the president of one of AlliedSignal's European businesses had directed us to consolidate a German center that weekend. No wonder he appeared agitated. We had a test scheduled to start the next day, but we were not prepared to turn this test into a real data center move.

"Why the rush," I asked.

"Evidently, the Works Council has decided to file a protest Monday. If they do, they'll tie us up for months."

"That doesn't make sense," I said, "There are only three jobs involved and they've known about it for months."

"I know, but they've changed their minds and now they're committed to stopping it."

"Okay, let's get everyone out of the meeting and see if we can figure out how to get this done."

After we decided that it was possible to pull this off, I cautioned everyone to use the *Mission Impossible* approach. *Mission Impossible* always ends with the bad guys about to discover the good guys in the midst of executing their elaborate plan. When they become aware of imminent discovery, they don't quicken their pace, but instead move with even more deliberate care. The good guys know any error would now be fatal, so they avoid panicky moves that could cause a mistake. For the next three days, we dispersed people to five countries, improvised plans on the fly, forgot what our respective beds looked like, and completed our own impossible mission with steady precision.

Despite organized resistance, speed kept our technology initiative on track and ahead of budget. If you plan on a lot of enterprise technology initiatives, be prepared for some heavy resistance. Standards, common solutions, shared services, and "centers of scale" all dictate a certain way of doing business. People resist change, especially if it includes a loss of control they exercised in the past. In the Chapter 6, we'll discuss ways of overcoming resistance; right now I just want to point out that speed represents one of the best ways. The faster you move, the less time resistance will have to solidify and become a serious problem.

Benchmark Operations

Benchmarking is the best friend of those responsible for enterprise initiatives. You may react negatively to this overworked business fad, but benchmarking can reduce pain and suffering. Three types of benchmarking I've found useful are listed in Exhibit 5.4.

Benchmark existing cost and service levels before you start any new initiative to counter anecdotal evidence that the initiative has created problems. To the greatest extent possible, measure performance with the same metrics the business units used.

When we decided to upgrade site infrastructure, the CTC started with a detailed survey of every facility. We knew a problem existed but needed to define the scope before designing our program. A questionnaire asked for current spending, number of people employed, equipment installed, measurements of performance, processes in place, and lots of other details important to designing the initiative.

Now, for a little truth. People in the business units easily restrained their enthusiasm for programs of this type. I'm sure you've heard the two biggest lies in the world, "Hi, I'm from corporate and I'm here to

---

Exhibit 5.4
Benchmark

Operations prior to initiatives.
Period over period internal operations.
Other external businesses.

---

help," and, "Thanks, I'm glad to see you." You can't fling a questionnaire into the field and expect a timely and accurate response from people trying to take the next hill. Delays in responding to data requests and the submittal of poor or skewed information are among the many ways of fending off corporate interference. The planning team must be persistent and bullheaded to get accurate information—all the while treating everyone with style and grace.

The accounting discipline constantly measure one periods financial performance against previous periods. Do the same for technology initiatives, but also include customer satisfaction scores, service levels, problem occurrence, elapsed time for resolution, cost per unit of service, and any other operational detail that affects the customers perception of performance. Watch the trends and take the appropriate actions to continuously improve performance. A history of better and better service at constantly declining cost goes a long way to quieting the inevitable complaints from the people who believe they can do the job better locally.

Besides internal benchmarking, continuously measure yourself against other businesses. Within the computer technology field, services exist that have extensive data on cost and service levels across all of industry. The information helps improve performance by highlighting specific areas that need improvement. If done year over year, you'll receive timely notice when other business have made a major breakthrough you missed. When this happens with our benchmarking, we always find a technology or process improvement that we had overlooked by being too internally focused.

Don't fear poor comparisons in one area or another; nobody does everything galactic class. I present the benchmark results to the CTC's executive oversight committee. A financial officer once said, "As I understand it, the red highlighted areas are good news."

"No," I responded, "The green areas represent best-in-class and red means below industry standard. We need to work on the red areas to reduce costs."

"Right, red means we can continue to expect cost reductions. All green would signal the end of the gravy train."

I loved this guy, his glass was always half full.

Manage Supplier Relationships

Managing business technology means managing suppliers. Since we've all become so dependent on equipment and service suppliers, Chapter 11 discusses this crucial aspect of modern technology management. Right now, I just want to point out one of the supplier programs we used with great success at the CTC.

One of the best ways to improve the management of your suppliers is to reduce the number of them. Many manufacturers trim their suppliers down to a few strategic partners. AlliedSignal aggressively follows this strategy for its manufacturing operations and it works just as successfully with technology suppliers. Immediately after the consolidation of all 11 data centers, the CTC supported 540 products from 119 vendors—a legacy of dispersed procurement decisions. We reduced the number of suppliers in half and continue to whittle away at them.

We call our program "Gold, Lead, or Dead." It uses the typical three-tier system, ranking vendors on a quality rating similar to that used in manufacturing, including on-time delivery, quality, price, and after-market support. Gold suppliers have a right to bid on any new procurements and we give them the privilege of making technology briefings to the CTC staff. The Goldies receive an invitation to our annual suppliers conference where we lay out our plans for the following year. Lead suppliers may bid, but they must win overwhelmingly on technology, price, or both.

Dead suppliers do not receive requests for proposals (RFPs) and find their products actively reengineered out of the CTC.

We make a big production out of the suppliers who move between tiers. An announcement letter is sent to our salesperson, with copies distributed to every sales and marketing executive in the supplier company. Salespeople have told me that the list of Gold suppliers in the reception room has prompted discomforting questions from their marketing managers when they notice they are not on the list. This program has a surprising effect on vendors; with the promise of a little status and a few competitive perks they suddenly seem intent on moving up, or maintaining their position.

Overcome Resistance

Attractive business cases can be subverted by resistance to organizational change. To overcome inertia, you need to break through resistance to obtain executive approval and then to execute the program. Most technology initiatives that fail, succumb to organizational resistance, not technology flaws. The next chapter will present my prescription for managing change. The behavioral aspects of technology initiatives should never be underestimated. People naturally feel uncomfortable with change, especially when they perceive it as threatening their livelihood or limiting their control over the performance of their job.

Pining for the Good Ol' Days

"Are these the best systems you ever had?" I asked.

My question to the director of production control came over a pitcher of beer to celebrate a physical inventory that came in right on the money. I had just spent a year implementing state-of-the-art manufacturing systems for a Midwest division of a large corporation. My pride in the applications and our swift implementation prompted my cocky question. His response quickly deflated my ego.

"No," he said. "The best system I ever had was when I sat on a platform over-looking the assembly lines. I had a 10-foot status board behind me where I posted the jobs running and the ones queued up ready to go. Six runners checked stock levels and pulled assembly kits. Anyone could glance up and see the exact status of the entire factory. That was the best system I ever had."

"Wait a minute," I asked, "what were sales back then?"

"Oh, less than a fifth of what they are now. I know that method wouldn't work today, otherwise I'd still be using it. I had control back then and I'd go back in a heartbeat, if I could."

Further discussion proved he was serious. This explained a lot about his strong resistance to the new systems I had been brought in to implement. It definitely had been "show me" each step of the way. Technology initiatives cause massive organizational change. People fight things they don't like or understand, so you need to prepare for the backlash against your technology initiatives. The next chapter describes how to manage change so you can execute your initiatives without causing too much civil unrest.

CHAPTER 6

CHANGING ROOM

Manage Your Changing Environment

I tell you sir, the only safeguard of order and discipline in the modern world is a standardized worker with interchangeable parts. That would solve the entire problem of management.

—JEAN GIRAUDOUX

Nothing challenges our skills as much as change management. If you examine your agenda of things to do, you will likely find that you function primarily as an agent of change. That's the nature of management. Companies don't pay big dollars for us to act as caretakers for perfectly positioned organizations that exhibit superlative performance. Even if you were to have one of these jobs, nothing stays static, especially in the technology field.

Corporate technology initiatives always include a large element of organizational upheaval. They tend to centralize a service or control; force product or process decisions on people who would rather do their own thing; or cause a change-out of one technology for another. Worse, many run counter to the culture of the personal computer. The stalwarts of the information revolution don't like it when the centralized technology organizations co-opt their industry and inventiveness (cultural issues within computer cliques will be described in Chapter 9).

Directing massive change, especially when it has all the earmarks of a counterrevolution, whips up intense resistance.

Once, an executive asked me to change a briefing I had put together to kick off an initiative. My chart stated that one of the barriers to success would be resistance due to loss of control. He wanted me to change it to "perceived loss of control." I refused, explaining, "No, it's not a perception, it's real." If you don't want to be blindsided by unexpected resistance, then you need to face it squarely.

Executives frequently encounter frustration when they direct some change and later discover nothing happened. Every organization has bureaucratic tendencies that develop naturally as a protection from internal and external interference. Mere direction by even a forceful executive does not alter an organization's behavior patterns. If the executive persists, then the organization, as a ruse, frequently assumes the pretext of having adopted the change. The challenge is not to perceive the need for change nor even to define it, but to actually get it incorporated into the organization's culture and practices.

You cannot abdicate your responsibility for managing change by forming teams and telling them to go figure it out. All the propaganda about change is meaningless unless a leader's vision sets a true course for the future. Change for the sake of change is ludicrous. You don't change your spouse or home on a whim, and you shouldn't change your organization without good reason. Once everyone is shown the destination, understands the flight plan, and knows the assignment, then empower teams to take care of the thousands of details required to arrive on time with the ship intact.

Balancing Act

Some would say that the change theme has been worked to death. Personally, I now apply a half nelson to anyone I encounter telling me to "embrace change." This doesn't mean I don't believe in the need for change or that it doesn't represent the most important responsibility of management—only that we need to approach change with a balanced perspective.

I learned a lesson early in my career. As an applications designer, I frequently asked people why they did things in a certain way. The usual refrain was, "Because that's the way we always did it." I had bought into the common wisdom that you dismissed this answer and redesigned the process. Except I got burned badly once—the process I excluded from a new application had sound justification.

The person I interviewed didn't know the reasons for the process, but that didn't alter the fact that the process had validity. From that point on, whenever I encountered the "we always did it that way" answer, I added it to a list of candidates to eliminate, but did some further background checking to determine whether someone, somewhere in the past had known exactly how to design the procedure.

You need to understand exactly what you are doing when you change things that appear to be working. I still believe, "If it isn't broke, don't fix it." We've legitimately broadened the definition of "broke" to include excessive cost, unacceptable cycle times, unique processes for the ordinary, and organizations hiding in silos. My complaint is with those who just toss everything up in the air without a well-established strategy. Without a clear vision driving the outcome, things seldom settle back down with any semblance of order.

When the CTC implemented its enterprise network, or intranet, the strategy included challenging every existing practice. Without sufficient thought, a third-tier carrier was selected because the pricing looked attractive. Our telecommunications analysts tried to convince management to stay with a major supplier that provided service across all of North America. We learned a lesson. The selected supplier subcontracted service in some parts of the country and difficulties coordinating with multiple suppliers elongated outages. AlliedSignal returned to a major carrier, but not before unhappy customers made their displeasure known.

Mea Culpa

Despite my reservations about recent excesses in pointless change, I do have a confession. The material in this chapter comes from an article I published in the Fall 1985 issue of the *Journal of Information Systems Management,* "The MIS Executive as Change Agent." Yes, I'm one of those change advocates, but at least I led the herd.

This chapter provides some tips and practical advice for managing change, decreasing resistance, and generating support for technology initiatives. There are no grand revelations, only an attempt to articulate common sense actions that improve your ability to manage change.

Know Your Enemy

You need a plan of attack to overcome resistance, and the first element of planning requires knowing the enemy. Resistance or support for change is not absolute, but a continuum, as reflected in Exhibit 6.1, from a vested interest in driving the change to fighting change openly:

- *Vested Coalition* A vested coalition exists when the business and initiative managers believe the initiative to be a mutual responsibility and they both commit resources and their reputation to making it happen.
- *Supportive Coalition* A supportive coalition is characterized by the business saying, "We're 100 percent behind you. Let us know

---

Exhibit 6.1
The Supportive/Resistance Continuum

Vested coalition.

Supportive coalition.

Passive support.

Passive resistance.

Active resistance.

Confrontive resistance.

Polarize Resistance/Support

---

what support you need and we'll make sure you get it." They support the initiative, but believe it to be the sole responsibility of the initiative manager.

- *Passive Support* Passive support means the business does not oppose the initiative, but doesn't support it enough to get behind it. If it can be implemented without affecting business leaders' ongoing efforts, they don't object. They give verbal support in your presence, but omit giving directions to subordinates that communicate any sense of necessity or urgency.

- *Passive Resistance* Passive resistance occurs when an organization opposes the initiative, but momentum and political influence make open resistance futile or career threatening. Leaders send subtle signals to other organization members that they oppose the initiative, encouraging the absence of substantive action.

- *Active Resistance* Active resistance results when an organization opposes an initiative and believes it has an opportunity to stop or alter it. Opponents actively recruit allies and express their objections openly.

- *Confrontive Resistance* Confrontive resistance is characterized by someone euphemistically saying, "You do that, and I'll punch your lights out." The target organization feels severely threatened—functionally or personally. Whereas an active resistant coalition may be swayed by compromise, a confrontive coalition wants to kill the initiative.

Look at support and resistance as a gray scale. If you develop finely honed antennae for resistance, you can save a lot of aggravation by stomping out small fires of dissent before the wind picks up. Exhibit 6.2 lists what techniques are useful for dampening resistance.

Extremism . . . Is Not a Vice

You want to move up the Supportive/Resistance Continuum; however, my most important piece of advice is to polarize the situation. Passivity,

---

Exhibit 6.2
Fifteen Ways to Increase Support

1. Polarize support/resistance.
2. Secure, and retain, leadership backing.
3. Obtain public declarations of support.
4. Assess degree of change.
5. Demonstrate the need for the initiative.
6. Get business leaders to request the initiative.
7. Require businesses to participate in initiative design.
8. Communicate extensively.
9. Establish personal contact.
10. Straight talk with resistance leaders.
11. Educate extensively.
12. Reduce threats.
13. Time initiative for least business impact.
14. Move with speed and phase implementation.
15. Advertise and build off demonstrated success.

---

an insidious condition, sneaks up from behind and kills initiatives. Strongly supported initiatives succeed, and those with strong resistance never get attempted. Don't believe progress has been made by moving resistance from active or confrontive to a passive state. Unless it continues to move out of the passive state, the situation is worse. If support cannot be generated, then abandon the initiative or make the necessary adjustments to gain support. Your initiatives may be technical, but acceptance is behavioral.

When the CTC built AlliedSignal's router-based network, it did not start from scratch. A few of the progressive business units had already linked their sites together with routers and the CTC incorporated these into a North American grid. One business stalled, delayed, and postponed turning over their network to CTC control. They politely argued the merits of their architecture versus ours, respective cost structures, addressing schemes, technician skill levels, and any-thing else they could think up. Every meeting to resolve issues seemed to get rescheduled several times. Each time we tried to set another date, schedule conflicts arose that put the next meeting off for another four to six weeks. When the meeting finally did occur, they some-how forgot some vital piece of analysis they promised to remember next time.

We knew their behavior represented passive resistance, but we were distracted by other priorities and intended to deal with the situation later. This ended up being a mistake. One day the whole network crashed because the recalcitrant division made changes resulting in addressing and configuration conflicts. Events brought the issue to a head. If we had forced the situation, instead of acquiescing to their passive tactics, we could have avoided a significant business disruption. Worse, if the outage had never occurred, others might have been encouraged to follow their lead, jeopardizing the initiative.

Influence Peddling

My original thesis recommended securing top management backing. I had to modify this point after two discoveries: Other people had influ-ence beyond their title, and I lost my president's backing once on an im-portant initiative. Now, I suggest gaining the support of all opinion leaders—and retaining it.

The first person to gain the ear of a pivotal person has a distinct ad-vantage. Get in front of resistance building by establishing the proper groundwork to block resistance before it solidifies. You can dilute resis-tance by securing top management backing, but you need also to gain the support of every opinion leader. In the technology field, organizational attitudes and opinions frequently get set by a few individuals who have earned community respect for their technical prowess. Every organiza-tion also endures a few who relish usurping authority from those in lead-ership positions. Know who these people are and either win them over or marginalize them.

Yea, Verily

Once momentum has been established, create situations where supporters make public declarations that broadcast their support. This reinforces support and precludes opponents from underestimating the amount of political backing for the initiative.

In a former position, prior to joining AlliedSignal, I was the victim of the most Machiavellian example I ever saw of this technique. One day, our division president called me into his office. He started the conversation by describing the difficult performance we needed in the coming year to satisfy our corporate masters. He then elicited my support in making the difficult numbers. I agreed, of course, but then he asked an odd question: Would I shake his hand on it? When I answered yes, he said great, opened his office door, and ushered in the company photographer. You guessed it; the next staff meeting found the executive conference room adorned with all of our pictures shaking our "prince of a leader's" hand. The motto above the montage said, "Committed To Make It Happen." Needless to say, we had the best year in the company's history.

Mountains and Molehills

Every enterprise-wide technology initiative is different. You need to assess the degree of change for each and gauge the impact on the organization. Factors that influence acceptance include the degree of change from previous operations, the size and culture of the organization, the schedule for implementation, threat perception, and the organization's adaptiveness and tolerance for ambiguity. After you assess the degree of impact, you can estimate the time and energy necessary to build support for the initiative. Persistence is the key to success. If you lose a sense of urgency, then momentum recedes and the initiative gets smothered in the bureaucracy.

Our GlobalLink initiative assaulted the values and culture of the personal computer set, making it difficult to overcome resistance. Data center consolidation didn't cause many problems because AlliedSignal had been consolidating centers for years and this program only extended a trend. With the exception of the previously mentioned business unit, the enterprise networking initiative was welcomed by people because they didn't have the wherewithal to do it themselves.

In each case, we had to assess the degree of organizational impact and adjust our time, tactics, and energy to address the emotional response to the initiative. Every initiative affects a different part of the organization and elicits different levels of resistance or support. Don't make the mistake of approaching everything the same way or you may be blindsided.

Because I Told You So, That's Why

As a first priority, convince people they need to change. If people do not understand the need, you'll never gain their support. You may believe an enterprise initiative will enable network computing, improve performance, decrease outages, and lower cost, but if the troops in the field think they can handle it on their own, you will have an uphill battle.

Larry Bossidy has a unique ability to invent scenarios where change becomes a condition of survival. Through exceptionally aggressive schedules, organizational changes, budget reductions, stretch goals, or the insistence on a new business strategy, he creates situations where everyone recognizes the need to change. This continues to drive the company toward improved performance.

Whether the situation is forced or the result of natural events, your first step should be to explain why things need to change. People need to believe a rational business imperative drives an initiative, not empire building or ego gratification. So start every initiative with a communication program describing the reasons for the initiative and its value to the organization.

Make Them Think It's Their Idea

Initiatives always garner more support if the idea comes from the business. If you have a solution for a real business problem, then getting the business leaders to initiate the request should be easy.

Initially, the businesses fought central electronic data interchange (EDI) because they feared they couldn't control the performance of a critical business function. As a result, the CTC's EDI staff ran into lukewarm cooperation by the divisional analysts. However, after the EDI processing center proved the advantages of a centralized service, the business leaders became enthusiastic and pushed workload into the center as fast as we could absorb it. As confidence continued to build, the businesses proposed a supplier program on top of customer EDI. Because it was the business units' idea, the supplier initiative moved exceptionally fast with excellent site cooperation.

The GlobalLink initiative started with the business leadership, which helped overcome strident opposition from the personal computer crowd. Although the business leaders were indifferent to the initial data center consolidation program, once they saw the savings, they drove the CTC hard to consolidate any centers picked up through acquisitions. In every case, projects went smoother and faster when the businesses, rather than the CTC, proposed a initiative. When you see a need for a corporate technology initiative, start the process with a few private meetings with the business leaders to garner their support, and then try to get them to assume the leadership role in proposing the program.

Teaming Arrangement

Once you get the businesses to request a fix for a problem, don't run away and emerge months later with a comprehensive plan. Recruit locals to participate in the process. You'll end up with a better plan, and the business leaders don't resist things they helped design. At AlliedSignal, on approval for an initiative, client executives visit the various sites, setting up planning teams and getting ideas on how to proceed.

In Europe, this approach proved very valuable. Service suppliers rarely have pan-European coverage, and without in-country guidance, we would have wasted time looking for reliable contracted services.

Local knowledge also helps us avoid making legal, social, or cultural mistakes.

It's the Message, Stupid

Communicate early and often. Communicate with Web pages, site visits, newsletters, announcements, video and telephone conferences, and share all your plans and schedules. If you do all this, you still may be accused of not communicating enough, but don't give up, just keep repeating your message.

AlliedSignal always includes a comprehensive communications plan with every initiative. The plan includes hard schedules for road shows, publications, one-on-one visits with business leaders, presentations at company conferences, and anything else that helps get the message out. When an initiative is undergoing rapid development, it's easy to put off communicating unless a published schedule reminds people that these duties demand equal weight with other tasks. This intensive communication keeps everyone knowledgeable, but it has a second, less obvious, purpose. The constant need to report progress puts enormous pressure on the team to stay on schedule.

Up Close and Personal

Establish personal contact with people to give them a chance to judge the character and motivations of those directing the initiative and to build mutual understanding of objectives and issues. Don't succumb to the temptation to use impersonal e-mail for most of your communications. Get out there and meet people and let them get to know you. After a personal contact has been established, written communication becomes more valuable and effective.

On one of our initiatives, we had serious resistance problems from the local technicians. After a couple hours of private discussion with the site director, the two of us were able to clean up some misunderstandings, compromise on a few real issues, and build a long-lasting cooperative attitude between the CTC and the site. I firmly believe this working relationship would never have been established with a telephone conversation.

Resistance Forces

You can gauge resistance more easily from up close than from afar. Personal discussions with the site personnel provide an early warning system for trouble looming on the horizon. Only personal visits pick up subtle signals indicating passive resistance.

Your communication plan may need to include some frank discussions with resistance leaders. These are better done in person. Sometimes they work, other times they don't—I've had both experiences. Most of the time, I've been able to blunt an embryonic rebellion by having a private discussion with the ringleader. We did have to terminate one executive a few days prior to starting an initiative at his site because he would not stop attempting to undermine the program. At least my conscience was clear because we issued several due warning notices in advance of taking this unpleasant action.

Comfort Zone

Most people feel comfortable that they can continue to succeed by following their current practices. Lack of confidence causes resistance. Overcome this fear of the unknown with extensive education and ensure that knowledge is both comprehensive and distributed. The earlier the training occurs, the more likely resistance will be tempered.

If at all possible, pay for the training out of the initiative budget. When the CTC establishes a new standard, the cost estimate for switching out the installed base includes training. Prior to the changeover, the CTC funds sufficient training for field technicians that they feel comfortable with the technology. During training, vendors naturally promote their wares and tout the advantages of their products versus that of their competitors; so another benefit includes a not-too-subtle sell job on the standard selection.

Threat Analysis

When you encounter resistance, you must be perceptive enough to see the real reason for opposition. Resistance often results from threats other than those specifically defined by the target organization. If technical dissension represents a diversionary tactic, it's important to refocus the debate on the real issue or you will spin your wheels addressing something that cannot resolve the conflict. As the concerns are correctly identified, making compromises, providing education, or building safeguards into the initiative can protect legitimate concerns.

For example, endless technical bickering frequently hides fears about job security or loss of local control. Compromises and initiative design can address control issues, but many technology initiatives do result in reductions in force. This is another area where you need assistance from Human Resources. To blunt resistance, Human Resources must have a good job-posting system, outplacement services, an extensive career development program, and decent severance packages. A good open discussion about the real threats demonstrates that they are recognized and, at minimum, can be dealt with openly.

Timing Is Everything

Every operational activity has cycles, and choosing a low ebb for the incorporation of change reduces the risk of disrupting the business. Accounting applications present difficult problems with timing. The finance people never want an implementation during the closing cycle, but the close of one period is always the beginning of another and accounting systems need to be installed at the beginning of a period. Oh well, we've lived with that one for decades. Less obvious, but actually more important for infrastructure initiatives, are application projects within the businesses. If an applications team has committed to install a whiz-bang network computing system on a particular weekend, you don't want to be interfering with their LAN infrastructure at the same time. The CTC uses the Client Executives to coordinate with the site people to make sure schedules don't conflict.

Piecework

Break the total initiative into components and implement them incrementally. A phased approach helps schedule implementations, increases the organization's ability to absorb the change, allows adjustments to encountered problems, and provides a mechanism to build off demonstrated success. The ability to continuously show progress provides another advantage to a phased approach.

Phased implementations developed a bad reputation, especially for applications, because computer people kept moving functionality to some undefined future phase. Just because the concept was abused, does not make it a bad approach. No application project should exceed six months, otherwise you lose management's attention. If you accept this rule, then no option exists other than to break large endeavors into multiple phases.

Infrastructure initiatives naturally lend themselves to phased implementation. However, conflicts can occur on the sequence of deployment. GlobalLink saw a raging battle over whether to do some "quick-hit" fixes, returning later to do major repair, or visit each site once to do a bumper-to-bumper overhaul. The all-out approach won because we wanted to show a step-function improvement after the team left, we didn't have time to make multiple visits, and the budget wouldn't support a piecemeal approach. This meant sites at the back of the schedule had to suffer until their turn came up, but demands on personnel and money limitations left no other choice.

An enormous amount of time can be wasted debating an issue where no real alternatives exist. If no other option can realistically be employed, then quit arguing and get on with the task. Otherwise, you'll eventually pursue the same course, only the delayed start will delay completion.

Success Story

Nothing works like success. A phase plan for implementation allows you to demonstrate success and build confidence. Keep momentum by widely publishing progress and on-schedule performance to give the initiative an aura of success. When we started data center consolidation, nobody thought we could actually pull it off, especially by the committed schedules. So people who didn't like the program held back, waiting for us to make fools of ourselves. By the time it became evident that we were knocking centers off left and right, the initiative had too much momentum to protest.

Similar predictions of failure haunted the enterprise networking initiative. When the North America program went smoothly, the chant mutated to, "wait until Europe, then they'll fail." Not only does flawless execution quiet naysayers, but leaking these predictions to the deployment team can motivate Herculean performance.

Greater Than Sum of Its Parts

None of these actions or tips will individually reduce resistance and build support, but in my experience, collectively they greatly enhance the opportunity for success. Good managers control events to get a predetermined outcome. The objective should not be merely to implement a particular change, but actually to increase the organization's absorption rate for change. Eventually, an organization can become receptive to change and even anticipate it as normal and necessary.

In Theodore White's book *In Search of History*, he wrote, "Threading an idea into the slipstream of politics, then into government, then into history . . . is a craft which I have since come to consider the most important in the world."* The organizational leaders who learn how to accomplish this with regularity will be the successful executives of the twenty-first century.

Lesson Plan

Let's do a quick review. We started with strategy development, then addressed the ever-present problem of funding technology initiatives.

---

* Theodore H. White, *In Search of History: A Personal Adventure* (1978), New York: Harper & Row.

You next need a master plan comprising a series of aggressive technology initiatives designed to put in place a consistent enterprise infrastructure. Unexecuted strategies bring no value to the business, so you must simultaneously execute each step as you plan the next initiative. By the way, I included a light reproach directed at your own culpability in creating this fine mess we're in.

In the beginning, I told you there are no silver bullets. To be successful, you need to pay attention to two more areas; selecting the right technology and managing your computer people effectively. Let's start with the issue of selecting the right technology. This high-level review will include enough information so computers will no longer be an enigma.

7

FASHION STATEMENT

Manage Your Selection of Technology

When you're in the muck you can only see muck. If you somehow manage to float above it, you still see muck but you see it from a different perspective.

—DAVID CRONENBERG

Don't allow fashion to dictate the pattern and fabric of your technology initiatives. As you select hardware, software, firmware, groupware, and whateverware, leverage the near-term trends to maximize your return on investment. Striving to stay fashionable in the fast-paced computer field can put your credit limit at risk—haute couture always comes at a dear price. Prior to investing in an emerging technology, make sure it gives you a competitive advantage consistent with your business strategy. Occasionally, it makes sense to jump in early, but most of the time it merely means that someone coveted an enticing piece of electronic gadgetry.

Investigate emerging technology when it holds promise for your business or you expect it to mature prior to deployment. Boutique solutions can also enhance your ensemble. Other things can wait until prices drop. For example, no one doubts that desktop videoconferencing will soon become as ubiquitous as camcorders. The early buyers of this technology pay many times what the laggards will pony up, plus the mass market offerings will make the early versions look slow and clumsy. You can't wait every time for optimum pricing or your competitors will leave you in their dust. So you need to stop, think, and then decide which areas provide the best opportunities for you and your business.

Deploying technology that meets your business needs requires knowing the trends and not repeating the mistakes of the past. Can we forecast the trends? Certainly. The trends are listed in Exhibit 7.1. Computers continuously become smaller and faster, with ever-expanding memories, while simultaneously consuming less electricity. Smaller, less energy-hungry machines can be made portable. Programming becomes easier and faster. Systems increasingly integrate functions across an entire enterprise. Connecting different machines, whether within the same building or across the world, becomes easier,

---

Exhibit 7.1
Information Technology Trends

Faster processors.

Increased memory size.

Smaller packaging.

Reduced power consumption.

Portability.

Increased network speed and bandwidth.

Cooperative processing.

Accommodation of multiple protocols.

Improved applications development capabilities.

Increasingly integrated applications.

Digitalization.

Automated transaction entry.

Lagging software.

Economies of scale.

Rapid price decline.

---

while the speed of transmission increases exponentially. Everything will soon be digital. Software lags years behind hardware developments. Nearly everything provides economies of scale for centralized operations. Despite these startling improvements, the overriding and predominate trend is rapidly declining prices.

Impressive when you think about it. How would you like to compete against this performance in your industry? Want to be even more impressed? Nothing will slow these trends. Computer technology improves at this unbelievable pace because pent-up demand provides a ready market for entrepreneurs. Computers, networks, office equipment, personal devices, and factory automation will continue to get better and better, while prices continue to decline. Leveraging people, processes, and organization design with computers has never been easier or cheaper. Why doesn't this ring true to you? Two reasons, you keep raising the bar and you have a short memory.

Through the use of computers, engineers design new products inconceivable only a short time ago, Electronic Data Interchange (EDI) allows products to be ordered and delivered faster than it used to take to prepare a requisition, applications progressively integrate the entire enterprise, people electronically visit friends, exotic places, and information caches from their homes, telephones fit in your pocket, while terminals on nearly every corner issue money, dispense gasoline, and pay for groceries. Exciting new applications abound everywhere, but we believe progress has been minimal because so many old systems linger.

Corporate technology initiatives are about making it easier and cheaper to replace existing systems with new applications. Building a digital organization requires selecting systems and processes that leverage your business and preserve, to the greatest extent possible, existing investment in computers. You need to avoid mistakes in order to get this job done with economy and thrift. Poor choices in corporate technology standards are expensive because the enterprise breadth usually calls for major implementations of new hardware and software. Selecting the right technology is fraught with difficulty. You need to pick suppliers who will be around when you need them, provide step-function improvement in productivity, negotiate standards with technicians spread across the enterprise, and accommodate a hybrid environment that can support multiple styles of computing.

The Trend Is Your Friend

AlliedSignal uses the computer industry economic and technical trends to dictate each step in its technology upgrades. As discussed, AlliedSignal did not initiate a single, omnibus initiative to build a digital organization. The strategy called for a series of quick-paced initiatives, each leveraging previous accomplishments. The early, preparatory steps took cost out of existing operations to fund later efforts. Follow-on initiatives prepared the enterprise for network computing by laying down a consistent and reliable infrastructure.

AlliedSignal strives for world-class processes, and "contemporary" technology. The company emphasizes first developing world-class processes because they immediately impact profits and must precede automation. Contemporary means technology ready for deployment on a mass scale. AlliedSignal, as a global company, needs worldwide consistency in its technology base. The advantage of this approach is that it leverages technology investment to the greatest extent possible. The company pushes state-of-the-art only when it can drive down costs significantly, or provides a major competitive advantage. This hybrid strategy succeeds for AlliedSignal because it fits the company culture, maximizes return on investment, and controls risk.

To control risk, the selection of technology must be managed carefully. Common problems include underestimating the difficulty of setting corporate standards or not accommodating different styles of computing. Using the dynamics of the technology industry to your advantage when designing a strategy helps avoid missteps. Let's start with a quick review of the computer technology life cycle.

Fashion Model

Standing idly by while everyone marches past, or striving to be far in front of the crowd in every endeavor are both prescriptions for eventual trouble. You need to chart a course somewhere between these two extremes. Your own business needs and competitive positioning should drive your particular strategy. Exhibit 7.2 shows the normal life cycle for most computer technology. The maturity path for a distinct technology has predictable stages that can be used to make reasonable business decisions for adoption.

Born Innocent

Emergence, commodity, and obsolescence/absorption represent the three technology phases. During the emergence phase, prices start high and decline rapidly. Early emergence is characterized by euphoric anticipation, theoretical purity, the absence of a required infrastructure, and an infantile capability. During late emergence, prices decline, holes in the technology get filled, and people start to recognize the cost of needed infrastructure. Clamorous demands for industry standards also characterizes the emergence stage. Every vendor fights to differentiate their products, which frustrates the early adopters. The call for standards is inappropriate, but not dangerous: inappropriate because it violates the Natural Law of Computing.

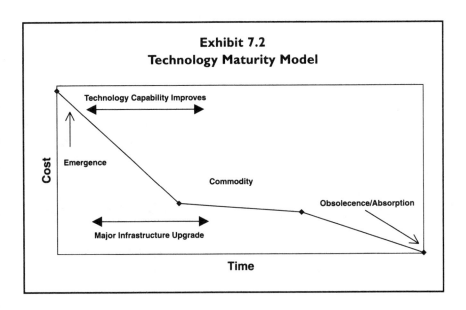

Exhibit 7.2
Technology Maturity Model

Law #5   Industry standards inhibit innovation.

Putting rules around something before it matures restricts innovation and slows further development of the technology. During the emergence stage, chaos normally reigns. People don't like the turmoil, so they try to put order into the world. Good objective, bad timing. The industry should wait until late in this stage before restricting the competitive forces of the marketplace. No real danger exists because the computer industry moves faster than the organizations charted to set standards. By the time the standards groups issue their rules, the technology has already matured to late emergence.

You only jump into early emergence when the technology can give you a march on your competition. Integrating any technology involves a learning curve and the early adopters do gain an advantage. It is fleeting, however, so most technology should be picked up in the later part of this stage. Let someone else exhaust their budget trying to figure out how to make a new technology work in the real world, but don't wait for the commodity phase. Industry leaders invest in new technology before optimum pricing, so if you dally too long, your competitors will be beating you with lower prices, better products, or both.

Seasoned to Perfection

An industry shakeout signals the start of the commodity stage, with mergers, acquisitions, and initial public offerings consolidating a few winners. Now everyone knows how to integrate it with everything else. You don't hear as much clamoring for standards because the winners have set the standard. Somebody may come out with a better permutation, but it is too late. Another law says the installed base dictates a de facto standard.

Law #6   Market share wins, not technical eloquence.

Once a technology has reached the commodity stage, prices have declined, you can hire people off the street who know how to make it work, and suppliers offer ready-to-wear solutions. It may be tempting to wait until a technology presents only minor difficulty, but you'll gain no competitive advantage when everyone can do the same thing as you. Ideally, you should adopt new technology in late emergence for your core business processes, and then rapidly propagate it to other processes during the commodity phase.

Grave Business

The obsolescence phase rings the death knell. Either something better comes along, or the functionality gets absorbed into another product. The latter predominates in software. Most bright software ideas start out as a separate product, but eventually die a lingering death as the functionality gets absorbed into some umbrella application. Examples include word processing software absorbing numerous unique accessories, MRP systems that include integrated accounting, and operating systems that imbed a Web browser. Hardware, on the other hand, normally gets quickly replaced with faster and smaller machines.

Should you replace everything that becomes obsolete? Eventually, but not necessarily right away. Just like equipment in your factories, lots of old technology keeps limping along because it hasn't reached the top of a priority list. The cumulative cost of replacing a large amount of obsolete equipment does present a danger. The actual cost of doing business can be disguised for a time by ignoring the need to refresh technology. If you let this get out of hand, your entire company may soon be in intensive care.

Setting Corporate Standards

During early emergence, several suppliers normally fight to establish market position. Since the technology evolves quickly at this stage, the suppliers rapid-fire new models, versions, or releases of their products into the marketplace. They also work hard to establish a large population of enthusiasts for their particular brand. This industry practice creates problems for corporate technology initiatives. By the time the need is recognized to establish a corporate standard, pockets of implementations have occurred in remote corners of the enterprise.

These small implementations harbor technicians who have done analytical evaluations of different offerings and invested time in learning to incorporate the technology into their installed base. The technicians supporting these implementations made their decisions at slightly different times, which means they evaluated the contesting products at different points in their life cycle. As a result, it is a happy coincidence when they pick the same product. More likely, you'll discover strong advocates for different products, causing standard selection to be frustrating, time consuming, and laden with emotion.

Enterprise solutions require that infrastructure design is centralized. If no central authority exists, then everybody feels free to investigate anything that warrants an advertisement in a trade magazine. Different individuals will frequently select different products as best-in-class. If you want enterprise network computing, then centralize decision authority to preclude uncontrolled cost, wasted time, and technicians fighting for their particular choice.

Taking a dictatorial approach alienates your computer professionals and large corporate standards boards quickly devolve into debating societies. AlliedSignal tried both extremes and settled on a small standards-setting forum. The company uses a lab at the Computing Technology Center to test new software and hardware. The lab technicians present the results and recommendations to the standards forum. In addition, the standards body frequently assigns project teams of site computer professionals to evaluate technology. The small standards body controls the process and acts as a clearing house for recommendations that go to the Executive Information Systems Council for final decision. (The council is chaired by the company's chief information officer.)

Despite AlliedSignal's efforts to design a smooth and efficient process, standards selection can still rankle those who strongly believe in a different standard. It helps to keep in mind that everyone may be right, but different perspectives lead to different conclusions. Corporate standards do sacrifice point optimization for consistency across the enterprise. One example, for global companies, would be the selection of a product with the best worldwide support structure, even though it may not be the best technical solution. Enterprise standards always entails compromises.

Here are two tips that reduce the arguing over standards: investigate all emerging technology immediately, and fund the replacement of non-standard equipment or software. If you let the trade journals and seminars tout a new technology without assigning an evaluation team, then an informal assessment will be made somewhere in the organization. Even if you're buried with other concerns, get someone working on the evaluation and publish the fact that a study is going on throughout the technical community. If you don't want to play catch-up, sanction official teams to evaluate anything new on the scene. When point implementations have occurred prior to setting a standard, then funding any necessary replacements from a corporate account eliminates most of the fighting. (If they knowingly violated established corporate standards, then insist that they fund the replacement.)

Architect a Multi-Use Infrastructure

Despite standards being difficult, most computer professionals can be easily convinced of their value. Sometimes no amount of logic can dissuade someone from a poor design approach. If you select a wrong standard, the consequences are normally sub-par performance for network computing applications. If you use the wrong overall design, the consequences will likely be an expensive failure.

Emotion—camouflaged with elaborate quantitative analysis—frequently drives design decisions. Resist the temptation to adopt a design approach that appears to make every problem go away. Paradigm shifts don't occur in information technology—unless you think constantly shifting paradigms makes sense. Although technology advances rapidly, the industry has sustained a predictable pattern despite its dizzying pace. Contrary to popular myth, no dramatic improvement in computing suddenly flashed onto the horizon to obsolete everything that preceded it.

The latest incantation says the Internet and World Wide Web will displace every other type of computing. Internet enthusiasts claim this technology antiquates all previous technology, destroys the Windows/Intel hegemony, and propels the information revolution to new heights. Perhaps, but client/server advocates made the same kind of claims, as did personal computer and mini-computer devotees before them. Instead of replacing existing technology, most advances in computer science end up increasing the number of ways we can design and automate business processes.

Granted, revolutionary products come to market every day. Hardware improves so fast, it plays havoc with our depreciation schedules. New software surprises and delights us with capabilities we never dreamed possible. Rapid obsolescence is an inherent characteristic of computer technology. The problem arises when an advancement in technology makes possible a new design architecture. An architecture is a style of building computer applications. Mainframes, personal computers, client/server, and Internet/Web all represent distinct architectures.

The euphoria that accompanies a new architecture causes an initial stampede to design everything using the new style. New architectures, by definition, combine machines and data in a unique manner. It takes time to discover that this new relationship has intrinsic strengths that make it more appropriate for a particular type of work, and that prior architectures still retain their advantages for other types of processes. Despite the rapid obsolescence of equipment and software, no design architecture has been supplanted by an omnibus, one-size-fits-all design style.

Building a digital organization without too many missteps, requires matching the appropriate architecture with the business process. NASCAR drivers don't bring a car from the showroom to the race track, and long-haul shippers do not use minivans. Commercial airlines use different equipment depending on the route, and military aircraft are useless for anything other than warfare. Similarly, you need to match your computer technology to the business process. In this chapter I'll review the major categories of computer architectures, and in the next I'll explain how to select the right technology from a business process perspective.

FASHION STATEMENT

Control Your Risk

Computer technology decisions must carefully consider the entire spectrum of available technology. Too often, the technology is selected first and then force fit to the problem. When a computer project flails endlessly, the root cause can frequently be traced to an early decision to use an inappropriate or immature technology. How do we get turned in the wrong direction? Project designers may recommend something they already know, avoiding new approaches that might be a better solution. Sometimes technicians push a new architecture they want to learn, so they can enhance the marketability of their skills. Other times, too much faith is placed in an over-exuberant sales pitch by a supplier.

The greatest risk of adopting technology too early, or too broadly, is disillusionment. A couple big ticket failures makes executives reticent about further investments in leading-edge technologies, while competitors, who held back or only experimented to gain expertise, gain competitive advantage with a better timed incorporation of the technology into their business. Witness the pull-back of many companies that went overboard trying to simultaneously convert all their systems to client/server. When something new comes along, over-enthusiasm reigns for a period and people conveniently forget about the prerequisite infrastructure necessary to make the technology function smoothly in a production environment. Eventually, unbridled passion recedes and the technology becomes part of our standard repertoire.

Knowing when, and how much, to invest in leading edge technology requires a business perspective missing from most computer decisions. You need to take a reasoned approach that balances risk with potential payback. Client/server and Internet/Web technologies can provide viable platforms for enterprise applications. Both, however, require an appropriate infrastructure, adequate support processes, and design expertise. These capabilities must be included in the business case or unwelcome surprises may cause the loss of executive support. Building a digital organization requires leveraging existing investments, using established technology for routine applications, and pushing new technology for completely restructured business processes that can make significant contributions to profitability.

The four major architectures include: mainframes, client/server, Internet/Web, and scientific computing. The strengths and weaknesses of each architecture came from the type of processes it was originally invented to support. Each capably handles its specialty, but supports other types of processes less adequately. First, I'll review the architectures, and in Chapter 8 I'll relate them to business processes.

Big Iron

Mainframes grew up in the transaction processing environment. The operating system, hardware architecture, and applications methodology handle massive transaction volumes coming in from hundreds, perhaps thousands, of places simultaneously. Mainframe computers should be called the Clydesdales of the computing world. You've probably heard them referred to as the dinosaurs. Not quite. Actually, mainframes have become a commodity, but they are not obsolete.

IBM introduced the System 360 mainframe in 1964. The machine included a hierarchical, modular design for both the hardware and the operating system. This new architecture boosted processing power and allowed IBM to offer a relatively granular and compatible series of machines. This breakthrough design provided IBM with a very profitable product line. IBM's ability to incrementally introduce the 370, the ES9000 line, and CMOS microprocessor versions from the same architecture generated enormous revenues. Unfortunately, it sowed the seeds for IBM's troubles in the early nineties. The System 360 grabbed enormous market share, generated 70 percent margins, and got IBM into antitrust difficulty.

In the 15 years that Thomas J. Watson Jr. reigned as CEO (1956 to 1971), IBM sales grew tenfold. To squiggle out from under antitrust scrutiny, IBM gave away its trade secrets. IBM licensed its operating system and published its hardware specifications to other manufactures, so today you can buy an IBM, Amdahl, or Hitachi mainframe with absolute confidence that all of your systems will run without even the slightest modifications. (Recently, IBM was relieved of its antitrust constraints.)

Now for the really bad news. Commodities don't sell for 70 percent margins and companies lulled by these kinds of profits can't perform in a savagely competitive market. Huge margins and galloping sales can lead to obesity and arrogance. Mainframe prices have been collapsing for the past several years and all three manufactures chase their tails trying to get rid of costs faster than the decline in their revenue.

The mainframe still depends on a recovering IBM for design leadership. IBM can no longer manage the price/performance curve and as a result, prices free-fall. The mainframe has competition not only from plug-compatibles, but also from microprocessor-based machines. What's my prognosis? Too early to tell, but one thing is certain—the mainframe will be around for a long time because billions of lines of code exist that cannot run on anything else.

Three's a Crowd

Mainframes have now accumulated three decades of applications. The trade press continuously publishes articles about downsizing and the ascendancy of network computing. These articles invariably predict that the latest innovation will finally drive a spike through the mainframe's heart. Don't believe these prophecies. The vast majority of network computing development is computerizing processes different from the ones that already exist on mainframes. This will become increasingly clear in this discussion. Right now, let's look at what does run on mainframes.

The three types of work on these huge beasts of burden include online transaction processing (OLTP), batch work, and ad hoc computing. The first two belong on a mainframe, but the third needs to be moved off as soon as possible. OLTP is the industry term for high-volume transaction processing that can come from many different locations simultaneously, as with automatic teller machines, retail bar code scanners, or inventory transactions from multiple warehouses. Batch processing means the work is defined by files instead of individual terminal transactions. Ad hoc computing means a person does whatever he or she chooses with a small piece of the mainframe's resources. Mainframes perform the first two with exceptional efficiency, and any computer can do the third. The mainframe will be with us as we enter the twenty-first century because it excels at these brutish tasks and it would cost an enormous amount of money to convert them to a different kind of computer.

Successful Operation

Mainframes do a good job with batch and OLTP because of IBM's operating system. What's an operating system? An operating system consists of program code that manages the workload and handles mundane, repetitive tasks. Operating systems loosely correspond to human motor reflexes. You don't consciously think about directing your arm to reach out for that cup of coffee and application programmers do not need to tell a computer where to go to grab a piece of data.

IBM's current operating system can trace its lineage back to the first one introduced with the System 360. Through the years, it has evolved into an extremely sophisticated workhorse. The design approach does not give individual users access to every nook and cranny, but it does drive these expensive machines as fast and efficiently as possible.

The design focused on turning off the Wait Light. (Ancient computers, circa 1965, had a light on the operator console that signaled when the computer was waiting for something to do. Supervisors harangued operators to keep the Wait Light off. Although modern computers have no such light, nor do operators even see the computer itself, the doctrine of keeping the machine running at full throttle has become ingrained in the mainframe culture.) This bias toward continuous operation causes mainframes to run a heterogeneous workload very effectively with an operating system that drives the equipment hard.

Networking almost became the Achilles' heel of the mainframe. For a time, IBM stubbornly insisted on purging from their discourse any influence from another architecture, but a few years ago IBM relented, and now a mainframe can talk to any other computer. This business decision became crucial because the industry, led by client/server technology, was moving to network-centric computing.

Client/Server

The client/server architecture developed by meeting the needs of knowledge workers. From a hardware perspective, the architecture evolved from personal computers and local area networks. The basic concept calls for a client, normally a personal computer, to work in concert with a server. A consistent client/server attribute is the use of an intelligent device at the workplace, with a sharing of tasks between it and one or more servers. Client/server applications do not use the traditional mainframe architecture, which maintains a master/slave relationship between the host and a dumb terminal.

From the Desk Top to the Enterprise

I first encountered personal computers (PCs) in 1978 when I made a proposal to buy 20 for our accounting department. I finally got the 20 approved and immediately received an avalanche of requests for more PC stuff. Since that fateful day, the desire for more and better personal computing has never wavered. The PC has become as ubiquitous as the telephone, and just like the telephone, users want to be connected to everyone in the world. It didn't start this way. The industry called personal computers "personal" for a reason; they were an individual's private little device to do his or her own thing. Times change. The emotional attachment people feel for their desktop equipment is all that remains personal.

PCs started off as stand-alone computers, primarily used by people who worked with numbers. The PC spreadsheet became so prevalent that when I once asked our controller for a four-column old-fashioned paper type, he gave me a blank stare. After rummaging around every filing cabinet in the department, he finally gave me the only one he could find—a six column, perfect-bound, light green booklet with half the pages ripped out. Five years before, every number cruncher earning a salary had a desk full of a variety of these forms.

The primitive word processing on early models of computers took a user with an IQ of 140 to figure out the obscure combination of keys required to do the simplest task. Unwieldy software and dot matrix printers made office workers happy to leave these machines to the analytically inclined. The next generation of machines changed everything. In the early 1980s, new 16-bit monsters brought color monitors, laser printers, databases, and gargantuan memory sizes of 64 thousand bytes. Then the software houses started producing new applications and these machines became as desirable in the office as a VCR was in the home.

Reach Out and Touch Someone

Then an interesting thing happened—people wanted to share files with one another and they tired of passing around floppy disks. So the computer technicians started networking personal computers. Actually, the real impetus was the laser printer. It finally made a word processor more enticing than an IBM Selectric. The first laser printers were expensive, so if you wanted one, you had to learn to share. This started the demise of stand-alone machines. When the computer people tied these boxes together, they called them Local Area Networks (LANs). Early LANs were pretty small, primarily used to network workgroups and subdepartments (e.g., cost accounting, budgeting, pricing, manufacturing planning, marketing, financial analysts).

Engineers had their chance with the next generation of equipment. Luckily, generations in the PC world last only about two years. Nobody had to wait long. The fierce competition in the PC market drives innovation and invention at an unprecedented pace. These little boxes keep getting faster and faster with phenomenal growth in memory and disk storage. The software may lag by years, but that doesn't stop people from salivating every time a new model hits the showroom.

Ganglion Infestation

Soon we started networking small LANs into bigger LANs. During the mid-1980s, the raging debate argued the merits of different protocols and one brand of network management system versus another. Since no clear winner emerged, every conceivable type of network ended up being built. Ideally, all of your PCs in one location should be neatly tied together as one network or as a series of nicely bridged LANs. That is unlikely, however, unless you've made a concerted effort to standardize all of the disparate systems in your organization. If not, then you're probably encountering serious problems with the latest networking trend.

Everyone now wants to connect all the PCs and LANs into a single enterprise wide network, or intranet. Current technology easily facilitates interconnecting two remote sites, but the process exposes the incompatibilities within the sites. Each site proceeded along their own path with no central direction to encourage consistency.

Now most companies own a rat's nest of PCs, premise wiring, LAN managers, and assorted other paraphernalia. This presents a huge problem if you want them to work together as a single integrated computing network. Client/server computing brings new design options, but this technology carries with it a whole new set of automation problems. This is not technical problem, this is a money problem. If you spend the money to replace much of your installed base of equipment and software, then the technology can support an enterprise wide network.

Why might you want to do this? To support enterprise network computing applications. Common applications across the enterprise enforce common processes, reduce complexity, and improve organizational performance. Knowledge workers also drive demand because they've gotten used to trading spreadsheets, documents, and files with people in their own operation, and now they want the same capability with everyone in the enterprise and beyond. The emphasis on shared services, "centers of excellence," divisional realignments, acquisitions, benchmarking, and the encouragement of "best practices," makes common applications a critical element of a strategy to build a digital organization.

Common applications using a client/server and Internet/Web architecture need connectivity to every potential user. Since these applications are getting more ambitious, the connection needs to get broader. The benefits may appear nebulous at the moment because these are all first generation systems. It's going to solidify and get better very fast. That's the history of personal computing. Not too long from now, in fact, we'll probably be wondering how we ever got along without an enterprise wide network and all the applications that come with it.

Weaving a Web

The Internet started as an electronic publishing medium, with some rudimentary e-mail and groupware capabilities. The military and academia used the Internet to trade research, send notes, and collaborate electronically. Two technologies took the Internet from an interesting electronic backwater to the mass market. The advent of web browsers made the Internet easy and programming languages increased its breadth of products and services.

The Internet evolved from scientific computing and, without a web browser, it can be a daunting experience trying to figure out how to navigate the maze. The naked Internet challenges even experienced computer professionals. The Internet's arcane rituals served the original devotees fine, but it kept the number of patrons below the critical mass necessary to do business on the information superhighway. Just as graphical user interfaces made computers easy to use, web browsers disguised the complexity of the Internet with a simple, straightforward facade that appeals to everyone.

The World Wide Web experienced unprecedented growth, even for the lightening fast computer field. Everyone seems to be taking a test drive on the I-way, finding it a fascinating place to hang around. Now businesses eagerly build roadside retail outlets to take advantage of all the traffic. The problem is getting the traveler to stop and buy.

To put enticing products on the Internet requires building material. This came with the new programming languages—the most predominant being Java—that make it possible to go beyond publishing to all kinds of new and exciting applications. The most popular new systems allow customers to directly enter transactions. This class of applications provides customer convenience, reduces business cost, and increases sales. Other applications that dazzle the mind will follow soon. The Internet takes electronic commerce beyond electronic data interchange to paperless, visitless, and hassle-free customer relations.

Despite the euphoria, the Internet does have limitations. I'm not talking about the soon-to-be resolved issues with security, but more fundamental limitations inherent within the architecture. The Internet's strength still lies with the publication and distribution of information. The Internet efficiently transfers big chunks of data, but it is not designed for high volume transactional activity. Real, collaborative work by people in your organization needs to be kept inside your own electronic fence. The sheer size makes finding what you want tedious. Traffic patterns on the Internet have the same problems as real highways. The primitive tools to interface with existing applications reduce performance and design options. Last, and most important, no one manages this thing.

The most obvious way to overcome many of these problems is to use the technology on your own private network. This popular solution has caused a tremendous corporate interest in intranets, with off ramps to the Internet. But before you direct all your internal applications development to web technology, remember this architecture still needs operating system capabilities, middleware, development environments, data bases, transaction engines, testing capabilities, operational tools, and many other things we take for granted in other, more mature, technologies. I'm not suggesting you ignore web technology. You will do so at your own peril! I am saying you should use common sense and use the architecture when its strengths fit the application.

Scientific Computing

If your company builds products designed by your own engineers, then you have yet another set of computer technicians within your organization. This recalcitrant breed includes the engineers who, up until recently, kept themselves somewhat isolated from the other types of computing. Corporate initiatives—like concurrent engineering—now force them to integrate their systems with the rest of the company.

Scientific computing is different. Typically, business computing is I/O bound and engineering and scientific computing is compute-bound. An I/O-bound job does extensive input and output to external storage devices. Compute-bound means a job does very extensive calculations and the speed of the computer determines how fast the job finishes. Engineers like fast machines and business computers need very quick access to external files. Fortunately, business and scientific computers come together with the Internet. Since the Internet's heritage is scientific computing, integrating these two different architectures becomes easier every day.

Engineers and scientist push for faster and faster machines, especially high-powered workstations. These workstations tend to be connected so that files can be shipped between computers or to expensive peripherals like plotters. Networking becomes an increasingly important issue when work needs to be done by geographically dispersed professionals. Several companies have started to establish centers of excellence for specialties like metallurgy and stress analysis. This creates a demand to electronically ship large files between sites. Voluminous engineering data requires massive bandwidth; jargon for the capacity of networks and telecommunications.

Engineering represents a type of ad hoc computing you should move off your mainframes and onto workstations. Budgets, not technology, cause the slow migration. Data center billing systems charge engineers based on usage and these costs hit departmental budgets as an expense. Purchasing workstations requires capital budget. During capital planning, the engineers ask for piles of workstations. When a directive comes down to cut the capital budget, the engineers, faced with a choice between new paraphernalia for the labs or cutting the number of work stations, opt for the lab equipment. When push comes to shove, they can still do the work on the mainframe. If you want to accommodate both needs, then start leasing workstations or increase the capital budget. Simple.

Family Jewels

Software for engineering and scientific computing comes in three flavors: (1) purchased analytical, (2) computer aided design (CAD), and (3) proprietary. This software makes your scientific computers worth having around. The purchased analytical software is a commodity, and all your competitors have exactly the same thing. The CAD software comes in many varieties, but the main differentiator between you and your competitors has nothing to do with the specific brand. The proficiency of your use and the execution of a design strategy set you apart. CAD software can be a handy tool for your designers or a strategic linchpin in achieving your speed-to-market imperative. It all depends on your effectiveness in using standard parts libraries, judicious selection of two-dimensional, three-dimensional, and solids modeling, integration of analytical tools, networking, and the training of your design staff.

Your proprietary software provides the real added value to your product line. If you think you have something unique in your product offering, then try rummaging around your engineering department. I'll bet you find some proprietary software written by your own engineers that provides the key to this uniqueness. Despite the casual attitude most companies display toward this code, it is extremely valuable and possibly represents decades of accumulated design knowledge. I suggest you treat it as one of your most valuable trade secrets.

Open Minds

One last subject before we close this chapter. Open systems are a myth. The open systems concept represented wishful thinking on the part of customers who wanted to pick and choose from multiple vendors. Any piece of equipment would "plug and play" with any combination of other equipment. Now think about this a minute. In your business, would you like your customers to have the freedom to toss out your products for those of your competitors? Even worse, if this ever did come about, how much would you spend on research and development to keep your gear interchangeable with everyone else's? Vendors paid lip service to open systems while they continued to differentiate their products to the best of their ability.

In the past couple of years, suppliers have surreptitiously redefined open systems to mean common interfaces, not portability. Big difference—it still allows innovation and differentiation. Open systems, as originally conceived, was a poor idea. Why would you want to put a stranglehold on the most inventive and innovative industry in the world? Again, Internet/Web technology will come to the rescue. Standardized interfaces defined by the Internet/Web architecture allow almost any box to be attached to a network.

The open systems debate has receded, and attention has shifted to the implications of the Internet to client/server. The computer field has a tendency to hotly debate architecture, forgetting the purpose of computer technology. We as managers, however, must keep the focus on the application of computer science to our businesses. That means analyzing the process first, and then selecting the computer technology to automate the process.

We hire different skills for different jobs, because everyone within your organizations does not do their assigned tasks with the same processes, nor from the same knowledge base. Computer systems need to leverage these skills and knowledge to maximize productivity. Matching the computer architecture with the business process makes this a lot more likely.

CHAPTER 8

WORKING CLASS

Manage Your Business Processes

One machine can do the work of fifty ordinary men.
No machine can do the work of one extraordinary man.

—ELBERT HUBBARD

Work comes in many varieties. Some of us do, some of us think, some of us decide, while others merely fake it. Like work, computers come in many styles and flavors. Manufacturers spread tiny, fixed-function processors throughout your cars and appliances. You may own a laptop that does most of the functions of your desktop model and your company probably uses large computers to serve many people. Applications also come in many varieties. If you're lucky, your computer helps you to do your job with effortless ease. Most people are not so lucky. Computers can be ornery. Why? Because the people who design these things don't know how you work. People unrealistically expect system designers to know the intricacies of every job. Not likely.

We're going to look at the world of work and how computers can make life easier for ourselves and our customers. This will be a high-level discussion. We'll look at production processes, knowledge work, and decision making. Since we all work to please our customers—and attract new ones—we'll also examine interacting with customers. In fact, since customers are so important to our paycheck, we'll start with the ways computers can improve an organization's customer interface.

Make It Easy for Customers to Do Business with You

While attending a conference in the early 1980s, I met a bank vice president. After discussing several computer subjects, I asked her what she did at the bank. She said her responsibilities spanned all the aspects of the emerging market for "home banking."

"That's great," I said, "I can't wait to bank from home."

She looked pleased and said she was grateful to hear that, because she worried that people didn't really want to bank from home. She asked if I already had a computer in my house.

"Of course. I've got an IBM PC and I can't wait to slip my paycheck in those little slots in front so I can make my deposits."

Now she looked a little less pleased. "You can't do that."

"Oh no—well at least I can get some twenty-dollar bills shooting out those slots."

Now she was definitely piqued. "Oh no," she said, " you can't do that either."

"Then what can I do?" I asked.

"You can pay your bills through your computer."

"Why would I want to do that? How can anything be easier than writing a check? I don't see the advantage."

"Oh no, this is what I was afraid of."

The next time I saw her, she had changed assignments. Smart woman. Customers didn't want to sit down at a computer and enter transactions to pay bills. Remember, in the early 1980s, you rarely found a computer in the home and only dedicated hobbyists took the time to learn how to use those early models. The banking industry decided to breed automatic teller machines (ATMs) as an alternative. These clever machines did what the people really wanted: took their deposits and gave them cash on demand. The ATM met customers' needs and satisfied their passion for convenience.

Getting Customers to Do the Work

Getting customers to do your work can be profitable and at the same time make your customers happy. Does this sound like a contradiction? Every time someone uses an ATM, the person does work that used to be done by an employee of the bank. Yet, people do it eagerly. That's a nifty trick: getting your customers to do things previously done by people on your payroll. To pull this off, you don't need to spew money out of machines on every corner. Actually, smart companies shift transactional responsibility to the customer all the time. Successes in this area include the car-rental-return machines, pay-at-the-pump gas islands, Federal Express's customer miniterminals, Internet applications, and those kiosks with the touch screens that answer questions.

Some of this new gadgetry may frustrate customers. For example, have you recently tried to call a big organization for help only to have a machine answer your call? They call this device an Integrated Voice Response Unit (IVRU). Designed properly, these devices can quickly resolve customer issues and put a wealth of information at the caller's fingertips. Poorly designed IVRU applications provoke customers with long-winded directions to touch various keys that can drag a caller into an electronic labyrinth. People don't like getting the runaround— whether from people or machines. Remember, customers like ATMs because they give users convenience in return for the work they do. Don't let your IVRUs return irritation. There is a lesson here: You can save money by shifting work out to your customers; but, if you want happy customers, you need to make it worth their while.

Will companies continue to invent ingenious new applications that encourage customers to do their own transactional activity? Sure. Most of the successes in this field come from one of two approaches: either special devices designed for a particular type of customer transaction, or Internet applications that allow people to do business remotely.

The common characteristic of specialized terminals is that they reside where customers most want to use a company's service. Credit card readers at gas pumps save you a trip or two to the cashier and ATMs dispense cash where you're likely to need it. So examine your customers' habits and ask yourself when they most need your products. Next, figure out a way to immediately deliver your product or service on the spot.

The second popular technique uses a Web browser to provide a common interface for customer transactions. Business on the Internet started slowly, but it's picking up momentum as credit card security improves and people develop the habit of transacting business using this medium. Today, customers on the Internet can peruse (e)mail-order catalogs, check their account or order status, buy new software, download patches for software bugs, be their own travel agent, or access and fill out government forms.

Federal Express already uses this technology to supplement their single-purpose terminals distributed to large customers. Governments have been very successful in saving citizens a trip downtown to pick up some form or other. The city of Phoenix has a Web page that shows pictures of various government buildings. You double-click on the building to get a map with parking instructions, or go directly to the form you need, fill it out, and submit it back online. It sure beats standing in line at the other kind of windows.

If getting your customers to do their own transactions appeals to you, then do some heavy creative thinking. You can shift your current costs to your customers if you figure out a way to lure them with convenience, savings, speed, or an increased choice of options (Exhibit 8.1). To be successful, you need to figure out what your customers think is a pain in the tush, and then design something really simple to make it go away.

Exhibit 8.1
Shift Transactional Costs to Customers

Lure them with:
- Convenience.
- Savings.
- Speed.
- Increased choice of options.

Place Device at Point of Transaction

Is this easy to do? Of course not, or someone would already be doing it and enticing your customers away with their new gizmo. Anything that makes a big difference in the bottom line doesn't come easy. Otherwise, your competitors would be stealing your market share.

Aim for something that permanently shifts cost out to your customers. This provides a long-term financial advantage, even if competitors replicate your clever idea. When your competition can easily copy your inspiration, you should not expect long-term competitive advantage. Why? Because it's the Law:

Law #7   Competitive advantage is hard to gain and maintain.

Sounds simple, doesn't it? People often forget this basic truth. If your competitor can easily duplicate what you've done, then you'll find competitive advantage fleeting. The things that give real, long-term competitive advantage possess the common characteristic of being difficult. Money isn't even a barrier. Look at ATMs. Despite the cost to distribute these machines, once they became popular, everyone got into the act.

AlliedSignal's Approach

AlliedSignal uses two approaches to improve the customer-interface process. Because the Bendix and Garrett predecessor companies grew divisions with overlapping and confusing product lines, many customers had trouble navigating through AlliedSignal's maze of businesses. Where should they send a Request for Proposal, which division builds a particular part, who can answer a technical question about an AlliedSignal product? A customer might know that AlliedSignal competes in a particular business, but would have no idea which site to contact. To resolve this problem, AlliedSignal borrowed one idea from General Electric and invented another. The company built a Customer Access Center (CAC) and implemented what we call an "electronic storefront."

Each business had historically developed their own systems, creating a menagerie of applications. The company decided it couldn't wait for them all to be replaced with common systems, so an electronic storefront was implemented to act as a window into the business units' computer systems. The design of this client/server application occurred after improvements were made in the process and organizational structure. During the first phase, customer service representatives use the system to check inventory, order parts, issue shipping instructions, and do many other tasks. The electronic storefront has the same "look and feel" for every transaction and the customer service representative doesn't care that the system communicates with many different brands and types of computers. Phase two includes implementing a common sales and aftermarket system, as well as additional process refinement. The final phase will allow customers to directly interface with a simple and standardized system.

The Customer Access Center relies on skilled people more than high technology. A casual observer might mistakenly assume that the CAC comprises nothing more than a few telephone operators. The staff representatives do answer the phone and direct calls, but their talent for problem solving justifies their attractive pay. Whenever a customer contacts the wrong business unit, the call is transferred to the CAC. The CAC analyst, through experience, can frequently direct the call immediately. When a new query comes up, they use a simple database that relates products to business units and contact names. If the database provides no clear answer, then the analyst asks for a return phone number and starts a Sherlock Holmes routine. Normally, their skill allows them to return calls with the correct information within a half hour.

Does the Customer Access Center work? You bet. I met an engineer who asked if I had responsibility for the CAC. I told him we provide technical support, but the CAC reports to marketing. He then told me he had received a call from a customer looking for a valve. The customer, who had obtained the engineer's name and number from the Customer Access Center, wanted four valves and didn't balk at a price carrying a high margin.

I asked the engineer, "Were you the right person to contact?"

"Yeah, I designed the valve ten years ago. What I can't figure out is how the CAC found me. I knew exactly what the customer was looking for and where to find some in stock."

This sale obviously didn't move our stock price, but we could have easily missed it, or given a customer the runaround. I never did ask the CAC how they found this particular engineer. We provide them with a slick little system that allows them to build bread-crumb trails as they venture around the company, but I doubt it led them to the answer this time. Despite computer assistance, the telephone, with some shrewd investigative skills, proves to be their most valuable piece of technology.

Another alternative may have been an World Wide Web home page. Home pages are popular, high-tech, and sometimes abused. You may not know that many of the hits on your home page will be by your competitors. (Search firms also surf the Internet looking for names attached to in-demand skills.) Because your competitors will be nosing about, you need to be careful what you put on the Information Superhighway. Assume everything on your home page is as public as an advertisement in *The Wall Street Journal.* Sure, AlliedSignal could have added a company directory to its home page, but it would not have solved our customer's problem of finding the right person nearly as well as the Customer Access Center.

I've mentioned before that AlliedSignal's strategy may not be a perfect fit for your business. Our customers are primarily other companies, not consumers. Again, you need to develop your own strategy for making your customer-interface process better than that of your competitors. My recommendation is to pay very close attention to this area. New technology comes to market daily. Constantly ask yourself how new developments can be used to make it easier for customers to do business with you. If you don't, then one day you'll wake up to discover your competitors have stolen your most precious asset.

Production Processes

People on your payroll do the work that generates your accounts receivable. They're making product, pulling inventory, ringing up the cash register, shipping and receiving, handling material, inspecting, closing sales, opening accounts, designing new products, and doing all sorts of other tasks that eventually lead to cash in the till. Incidental to their job, they need to record these events in a computer. Everything that happens in your company ends up being recorded into a computer by someone. You didn't hire these people to do data entry; in fact, it frequently detracts them from their primary job. I call these people the Event Recorders. You want to let them get on with their work with as little interference as possible. Luckily, the technology is available to do precisely that.

In technical parlance, we call the capturing of events throughout a large organization online transaction processing, or OLTP. The invention of dumb terminals and supporting system software in the early 1970s provided the technology to automate this activity. New systems started shifting transaction reporting away from your data processing people to your line workers. Now, you need to adopt better ways to accomplish the same purpose, without those old terminals that siphon productivity from your employees.

Despite their lack of eloquence, terminal applications did the job for many years and probably still drive most of your transactional activity. These are the culprits derisively referred to as legacy systems. Most of these programs have not aged gracefully, and you may think you need to replace them all. Perhaps not. Redesigning the front and back ends of these systems can not only prolong their life, but make them as modern and easy to use as freshly designed applications.

Look Ma, No Hands

The business requirement to record events will not disappear, nor is a network computing application always the answer to making these applications easy to use. To provide the right answer, we need to frame the question properly. What is the proper question? How do we let these people get on with their work and eliminate the need for them to input data to computer systems? This means we want the event recording to be effortless, or nearly so. New and developing technology, as shown in Exhibit 8.2, makes this objective more of a reality every day.

When you go to the supermarket, the checkout clerk merely passes your purchases over an optical bar code reader and the machine captures the transaction. Bar code technology has become ubiquitous. Bar

---

Exhibit 8.2
Reengineering the Front End

Electronic data interchange.

Special, single-purpose terminals.

Control unit interfaces.

Bar codes.

Radio frequency.

Combinations of these technologies.

Event Recording Needs to Be Effortless

---

codes now litter our factories, products, and documents. You can even use bar codes to program your VCR at home. Passing a wand over a bar code doesn't require typing skills, barely interferes with the process, and never makes a mistake.

Bar codes don't work for all event recording, so other technologies came to market. Control units (actually small computers) now run our machine tools, warehouses, and material-handling equipment. The initial versions of these control units kept pretty much to themselves. They knew what was going on, they just wouldn't tell other machines. Things have changed. Today, most of these devices communicate their activities to other computers relieving people from having to record the events. Control units—sometimes combined with other technologies— measure the status of equipment, inspect parts, keep track of inventory, adjust assembly line speed, and even automate previously manual processes.

Radio frequency (RF) units represent another fast developing technology. RF units send radio signals from control units and handheld devices directly to computers. These gadgets will soon allow you to fly through highway toll booths and receive a monthly bill. You can buy shop gauges with built-in RF capabilities that automatically transmit inspection results to a statistical process control (SPC) system.

Handheld inventory devices read bar codes off bins, accept counts, and automatically radio the transaction to a computer. What will they think of next?

Making someone else do the recording, that's what. Moving the transactional responsibility to the customer, discussed earlier in this chapter, actually, is a long-running trend. In the late 1960s to early 1970s, companies had rooms full of keypunchers and data entry clerks. Few are left. The theory says that to get accurate and timely data, you need to push the responsibility out to the person with firsthand knowledge of the information. So instead of sending a hard copy to data processing for keypunching, a terminal plopped onto someone's desk and the person was told to enter the transaction while doing the job. Thus were born online mainframe applications.

The theory was good, it just needs to be extended out from between the four walls of your business and into your customers' hands. They know how to accurately reflect the transaction. How do you do this without putting terminals in every corner of the globe? Easy, with Electronic Data Interchange (EDI). Industry councils define standard formats to electronically communicate business transactions. Actually, customers first pushed this technology, especially in the automotive business. They tired of lead times being extended by days, or even weeks, until someone finally got around to entering the order into a computer.

Let's look at the way EDI improves the process between two companies. One company wants to buy something from another. The buyer has all the data in a computer necessary to prepare a purchase order. If the buyer mails or faxes the purchase order, the seller must reenter all the data into the seller's own computer. With EDI, an electronic message not only goes with the speed of a fax, but it doesn't wait for human intervention to start the delivery process in the seller's system.

EDI goes well beyond the simple explanation in the prior paragraph. Standards exist for almost every type of intercompany transaction, so look at all your trading partner activity and try to eliminate manual processes wherever possible. Don't restrict your EDI program to customers either; insist that your own suppliers also do business with you using EDI. Remember the old axiom; Prior to improving the efficiency of a process, determine if the process is necessary in the first place.

I suggest you also examine your intracompany transactional activity. If your company does a lot of internal business between divisions, then you may have difficulty communicating between computer systems developed independently. EDI was invented to resolve this problem between companies, and it works just as well for internal transaction activity. Many development projects include large budget requests to build "bridges" between the new application and legacy systems. Next time you see one of these, ask the project team if they evaluated using EDI instead of a bridge.

Effortless Power

Things happen all over your company; it's why you employ all those people. You need these events recorded, and your people have lost patience with antiquated systems. But don't run out and design new network computing applications until you determine whether you can use other technology to eliminate the process. If you don't have a human interface, you don't need a graphical user interface. If possible, move the data recording responsibility out to your customer. Otherwise, try to automate the recording with bar codes, magnetic readers, direct communication from control units, or wireless signals from RF units, or eliminate the process entirely with EDI. What computer do you use? Who cares. It depends on the specific application. If you're small, then a PC server may do. If you need to transfer information to other applications in UNIX®, then a UNIX design may be appropriate. If you have input coming from all points of the compass, then nothing beats a mainframe. No one cares what computer you use when you make the event recording effortless.

Guess who does the largest number of mainframe online transactions in the world? Federal Express. The transactions come from those handheld bar code readers whisked across your packages. Federal Express deserves its reputation for high-tech answers to customer service and productivity. Not very many people understand that Federal Express judiciously balances mainframes, client/server, Internet, and a host of other technologies.

Remember the $1.3 trillion worth of program code that no one has the money to replace. Redesigning the front and back ends of legacy systems provides the best way I know to preserve existing investments. Thankfully, new and emerging technologies have the ability to extend the life of less critical applications, while we focus our design talent on higher priorities.

Is transactional data important? Very much so. If you look at the successful strategic information systems written up in the trade press, you'll find that they are all data, not process, oriented. Remember the Law: Competitive advantage is hard. Processing logic can be easily replicated, but collecting valuable data and presenting it in new and unique ways is difficult. So pay attention to the mundane task of event recording and include plans to make it as effortless and accurate as possible. Next, look at the data you own and ask yourself how it can be leveraged as information by using it in new and creative ways. Capturing data at the source, with rules to enforce accuracy, creates the raw material for lucrative mining operations.

Knowledge Work

The new data mining technologies allow analysts to research transactional data for operational trends and devise ways to use the information to improve sales. Another popular use of data mining is to reassemble the information so it has market value to external customers. If you need to figure out why your income statement looks anemic, when your customers are going to need replacement parts, or whether you hold the right inventory, then powerful new mining tools can quickly rummage through transactional activity to find the answers to these and many more questions.

Data mining represents only one of many proliferating technologies designed to meet the unique needs of knowledge workers. Other tools include e-mail, groupware, the Internet, data warehousing, statistical analysis software, presentation graphics, project planning software, laptop computers, videoconferencing, and cell phones. Software specifically designed for engineering, medicine, computer systems development, graphic design, and the financial industry are representative of the burgeoning market for knowledge workers. Additionally, most network computing applications address the needs of knowledge workers. The astounding array of technology listed in Exhibit 8.3 has come to fruition in the past few years, changing the way many people work.

In the beginning of this chapter, I said some of us earn a paycheck by thinking. Rather than knowledge workers, I prefer calling them "Data Sifters." Before knowledge can be applied, data must be transformed into relevant information. Data Sifters are becoming omnipresent. They uncover the whims of customers, provide insights into our business, invent new products, provide quantitative backup for decisions, and measure trends that lead to process improvement. These are very important people.

The rapid increase in power on the desktop, pent-up demand, and the transition from a manufacturing to a service economy drive a booming market in knowledge workers' tools. Actually, our economy is

---

Exhibit 8.3
Give Data Sifters the Right Tools

Data warehousing.

Data mining.

Groupware and e-mail.

Project planning applications.

"Same as, but" systems.

Custom applications.

Videoconferencing.

Office in a briefcase.

External networking.

Analysis and presentation tools.

Use Network Computing Technology

---

not transitioning from manufacturing to service, unless you measure an economy by the way people are employed, rather than by what the economy produces. After World War I, employment shifted from agriculture to manufacturing. We still produce more food than we can possibly eat, so the economy didn't switch directions, it expanded to new areas because technology changed the labor content of farming, while manufacturing technology increased the diversity of nonagricultural products.

Since World War II, the economy has been broadening at an accelerating pace as we decrease the labor content of manufacturing and increase the diversity of services. As more and more people make their living by providing or inventing services, they increasingly need network computing applications to improve productivity. The entrepreneurial computer industry has responded by introducing new products at an unbelievable pace. Let's look at a few of these.

Some Assembly Required

Data warehousing and data mining represent categories of the same technology. A data warehouse functions as a repository for the data, while mining technology provides the analytical tools. This concept is simple, but the execution is difficult. You move data from a computer dedicated to transactional work to another computer more appropriate for ad hoc queries. The difficulty and cost result from extracting messy data from multiple sources. Data warehousing projects must allocate a large portion of their effort to validating accuracy, adjusting data from different sources to the same format, building conversion tables, and accommodating decades of decentralized systems development.

The ideal solution would be consistent applications across the enterprise with standardized employee identification, account codes, part numbering schemes, location codes, and so on. Consistent data formats would also help, along with restricting people from putting information into inappropriate database locations. However, disparate data is a reality for most organizations, so make sure your warehousing projects include sufficient resources to clean up any data discrepancies. Since technologists don't eagerly embrace housekeeping assignments, be sure to include data analysts on your project team.

Leveraging Knowledge

Complex data sifting requires network computing and a host of related technologies. Knowledge workers tend to be collaborative, so groupware, e-mail, videoconferencing, and the Internet provide communication channels to their cohorts dispersed around the globe. As the Internet develops, it becomes increasing important to provide external networking so knowledge workers can tap into any number of databases out on the Information Superhighway. The knowledge worker's job entails extensive analysis, requiring a whole host of purchased and custom developed applications. Since they frequently don't make the ultimate decision, they need slick presentation graphics and other fancy tools to present their analysis to the decision makers.

Most client/server and Internet applications are designed for knowledge workers. One of the fastest growth areas for network computing is customer service. Every company wants to build elaborate customer service systems that put an unbelievable amount of information at the fingertips of the service representative who answers the phone. Since no one can anticipate every customer question, the customer service function needs access to numerous operational systems. The biggest difficulty for these applications is getting huge amounts of information organized into a single system. A network computing graphical user interface presents the information logically, provides quick transitions between systems, and reduces training costs.

To promote enterprise standards, implement applications that encourage "same-as-but" design. With this methodology, a designer must search a standard repository to find a part, financial instrument, graphic, software object, or list of features close to what is wanted and adjust it to a new form, fit, or function. This technique requires sifting through previously designed components and reusing elements that fit the new design. The absence of these types of systems perpetuates a lack of commonality within an enterprise.

A modern digital organization must also support employees who don't normally come on company premises to work. Provisioning a standard "office in a briefcase" to salespeople and traveling personnel presents a common face to the customer, increases productivity, and speeds communication between the office and the people sent into the field to find out what is really happening. To alleviate confusion and reduce training, applications for the road should be consistent with the products used in the office. The technology to support traveling employees can only partially reside in a briefcase. Many of the components required for seamless remote access to your organization's network must be installed as part of the on-site infrastructure. Ask your computer technical staff if they have provisioned the infrastructure for access away from the office or plant.

Who Are the Knowledge Workers?

The prior examples of knowledge worker applications may create a narrow impression of who does this type of work in your organization. My definition of a knowledge worker is anyone who accesses multiple repositories of data, adding value to the information with his or her expertise. This would include production planners, financial analysts, designers, planners, researchers, cost accountants, and academics, as well as people in publishing, advertising, medical diagnosis, customer service, and marketing.

Obviously, this is not a comprehensive list. Our economy has been growing knowledge workers like Tribbles in the famous *Star Trek* episode. These workers are also the ad hoc users of computers. As mentioned in Chapter 7, these people no longer belong on the mainframe. In fact, they never should have been there in the first place. The only reason they even ventured into the mainframe's maze was to get at the data buried deep within its recesses.

Knowledge workers feel particularly unhappy with the mainframers' perspective on computing. The mainframers' real problem was that their technology didn't allow them to meet the data sifters' needs. Traditional applications recorded events so individual queries could be made and batch jobs could use the data to schedule factories, print invoices, write checks, produce financial statements, or let customers know where their accounts stood. These hulking systems don't facilitate random searches for patterns. Mainframers didn't want to ignore knowledge workers, but their technology didn't lend itself to ad hoc requests and they were busy doing other things.

Why is this important? Because when network computing emerged, knowledge workers participated with others in creating the impression that the mainframe was dead. Mainframes still possess a superior architecture for heavy transactional workloads and production batch applications. Network computing has a much higher cost per transaction, and managing performance still presents a challenge. My advice is to ignore the overzealousness that temporarily comes with a new technology and use the most appropriate architecture for the job.

The Decision Makers

Computer science continues to make headway in technology that supports the decision process. Unfortunately, you seldom make the same decision twice, and computers do exactly as they are told, over and over again. It's not a good match. There have been some attempts to develop computer systems for decision makers, but their success has been limited by time or scope. Decision support systems (DSS), executive information systems (EIS), and expert systems (ES) all attempted to automate portions of the decision-making process (Exhibit 8.4). Some of them even work . . . for a while. But your attention keeps shifting, causing these systems to become outdated faster than day-old bread. This quarter it may be cash flow, next it may be sales growth, and the following year you may be concentrating on debt restructuring. It's a changing, dynamic world.

Exhibit 8.4
Decision Applications

Decision support systems.

Expert systems.

Artificial intelligence.

Executive information systems.

Invest Carefully

This doesn't mean you shouldn't pursue systems that assist in the decision-making process. Computers can provide decision guidance in repetitive situations. However, you need to understand the limitations of existing approaches, so let's look at some of the more popular ones.

Decision Support Systems

Decision support systems (DSS) became popular in the mid-1980s. About this time, I became involved with a DSS for Purchasing. AlliedSignal ran into problems with suppliers not meeting delivery schedules. During a site visit to expedite a slow vendor, we discovered something that we should have known all along: Suppliers accept every order sent to them, even if they don't have enough capacity to deliver on time. Few people have the courage to turn away business. After this less-than-startling revelation, we decided to refrain from putting into a shop business that exceeded 50 percent of the supplier's total capacity. That way we would not single-handedly overload capacity, nor be the cause of the company going out of business if we suddenly switched suppliers. We also decided to build a decision support system for the buyers.

The system needed to include something more helpful than rough statistics on business size. So we asked the buyers what they needed to make a purchasing decision. At the start, most people were skeptical, but soon the process began to take on a life of its own. To prove they could not be replaced by a machine, the purchasing agents bombarded the project team with endless lists of critical data and logic. They claimed they needed everything or the new system would be utterly worthless. The buyers' scheme, designed to kill the project with excessive demands, surprised us. We understood that targets of automation often feel threatened, but we didn't expect this from professionals. Usually clericals or shop workers suddenly develop an advanced state of insecurity with the proposal of a new system. Another Law states that projects grow in scope as customers try to squeeze in every possible piece of functionality, but the problem gets overwhelming when deliberate attempts are made to kill it with overload.

Law #8   The scope of every computer project grows.

You can't escape Law #8 for any system, but the phenomenon magnifies with applications for decision makers. It's the nature of the process being automated. Decisions absorb an unlimited amount of information. We succeeded in ratcheting the design back from decision making to providing some assistance in the decision process. Big difference.

Eventually, we included information on the size, capabilities, receiving inspection results, schedule performance, and buy history. The system worked wonderfully. When buyers received their requisitions, they searched by capabilities and got a listing of selected suppliers with their relevant histories. The computer didn't make a decision, it merely presented the options. I learned a valuable lesson; computers could be used in the decision-making process—as long as the decision making was repetitive.

Expert Systems

Expert systems fall into the same category; they work well in situations where things stay the same. I've seen good implementations of expert systems, but they have all been restricted to managing a process—a repetitive process. They can be valuable, especially when things go wrong in a predictable way. When something new happens, they can be a huge waste of time. People become reliant on the system and forget to think. Expertise comprises more than accumulated knowledge and experience, it includes an ability to think and logically work your way through to a new solution set. This skill will never be replaced by an inanimate computer. This doesn't mean that expert systems are not useful, only that they have limited use.

Executive Information Systems

Executive information systems are another story. Most of these end up being not much more than bulletin boards with a few extra features thrown in. They always start off well, but then the executive's interests change and the EIS can't keep up. I've seen a number of these systems with elaborate drill-down capabilities, but executives like Bossidy don't want to do the research themselves, they want the person who screwed up to come in and explain how he or she is going to fix it. Most executives who use computers find they can get along just fine with a spreadsheet, word processing, electronic mail, and possibly a Web browser. Executive-oriented software, with a few exceptions for repetitive situations, is not ready for prime time.

Computers do not, as yet, support decision making very well. Decision making is kind of fuzzy; it requires more than facts and rules. It requires judgment, especially high-level decision making. Computer systems started out handling the routine, repetitive events with fixed-format reporting. They quickly evolved from punched card systems, to batch systems, to online systems, to the automation of event recording. Network computing technology automates the knowledge worker functions by providing ad hoc access to data, effective personal computing, sharing of analysis and knowledge between team members, and ease of system navigation. The Law states:

> **Law #9 New computer technologies unveil additional layer of applications that suddenly become feasible and cost-effective.**

Every advance increases the portfolio of possibilities, and every advance brings with it a gaggle of gurus who claim the new technology will resolve all the problems in automating organizational processes. More likely, we will start automating additional processes. Since we don't always learn from experience, we go through the same predictable growing pains every time we try to make the latest technology work in the real world. These advances are heralded well in advance by technologists who have a theoretical perspective. They tend to promote the use of the technology with a purist attitude and inevitably claim the new advance has applicability to all types of systems.

If your job involves decision making, I have good news. Even if we cannot develop effective systems for you today, we are gaining ground. Don't get discouraged; dramatic new advances just around the corner will make your flavor of computer systems feasible. You're next in line.

Sizing Up the Situation

We have three types of workers: the doers, the thinkers, and the people who make decisions. A different combination of technology serves each best. Only it's not that simple—some people do all three kinds of work and normally you find all three in every organization. To make matters worse, you have to take into account different types and sizes of organizations (Exhibit 8.5).

Advances in computing technology often begin in service-oriented industries such as banking, insurance, entertainment, publishing, retailing, and brokerage houses. Because information provides the lifeblood for service companies, they tend to be creative in the use of these technologies. Their data centers equate to a manufacturing company's factory. Product development in these industries often comprises nothing more than a computer program that twists information in a new direction or delivers it in a unique fashion. Where they get this information depends on the type of services they offer. However they go about getting it, they want to collect it cheaply and then be able to

---

Exhibit 8.5
Technology Selection

Size of Organization:

Number of people.

Single locations, geographically dispersed, or global.

Type of Organization:

Service.

Manufacturing (process, repetitive, or job shop).

Type of Application:

Event recorders.	Doers.
Data sifters.	Knowledge workers.
Decision makers.	Deciders.
Hybrid.	Multiple.

Match the Technology to the Process and Organization

---

access it in unusual ways. The drive to remain competitive keeps them on the leading edge of computing and information technologies.

Manufacturing companies adopt new technology at a slower pace. They have a higher proportion of Event Recorders and their products have not historically been information based. Many learn to use customer records and aftermarket data creatively to generate new revenue, but for the most part, they limit their risky investments in computing technology. This doesn't mean they don't rely heavily on computers, merely that they tend to incorporate them after the technology matures. Manufacturers propagate new systems only after the technology demonstrates an ability to improve quality or reduce cost. Although they may not be leading edge, most manufacturing companies increasingly depend on information systems to design, build, ship, and distribute their wares.

Neither service nor manufacturing companies are homogeneous within these broad categories. Service companies come in unlimited variety, with many new ones soon to be invented. Each has something unique in the way it processes information. Manufacturing normally breaks down into process, repetitive, or job shop. AlliedSignal has three business sectors: Engineered Material, Automotive, and Aerospace. These sectors correspond with the three types of manufacturing. Each has its own set of problems and different priorities when it invests in computer-based technologies.

Finally, size must be taken into account. Everybody talks about enterprisewide systems, but an enterprise can be anything from a mom-and-pop shop to General Motors. Size brings people, and people bring complexity. When a salesperson calls and tells you he has an enterprisewide solution, ask for reference accounts from companies your size. Size even has a law associated with it:

> **Law #10   Size is the greatest determinant of implementation difficulty.**

If you belong to a large organization, especially a geographically dispersed one, then you need to use care in selecting computer technology. This doesn't mean that you shouldn't use a network computing design if you have 400 sites scattered across the globe. However, you should expect network computing applications to take longer, cost more, and initially be less reliable. This is not a value statement, merely fact. *Information Week* now rates network computing offerings by their ability to handle different-size organizations. This issue has grown to such importance that the term "scalability" joined the lexicon of the computer industry. Scalability refers to whether a network computing product can support an ever-increasing workload.

Every technology decision is situational. One size, truly, does not fit all. You look at the application, figure out what people need, and then choose the technology that meets the needs of the people who use it. In this chapter, I've tried to explain why your applications development people need a complete toolbox, unencumbered by any bias toward a single solution. To accomplish this, you need to bring together the specialists in each field so they can work as a team. It's impossible to win a basketball game with each of your players on a different court.

CHAPTER 9

DINOSAURS AND WHIPPERSNAPPERS

Manage Your People

Men will never be enslaved by machinery if the man tending the machine be paid enough.

—KAREL CAPEK

Some projects deliver on time, within budget, and include most of their promised features. Others slip and slide around and never reach their destination. What makes the difference? Winners effectively manage their computer professionals. There are two issues: the technology itself and the management of its delivery to the workplace. The first is inconsequential if you can't do the second. The delivery of good computer systems involves people, not technology. People choose the technology, people design it, people deploy it to the workplace, and people use it. Some managers don't fully understand computer technology, but they do know how to deal with people.

In the beginning, I told you that AlliedSignal's three tenets for deploying new computer technology included: (1) installing a technical infrastructure using aggressive corporate initiatives, (2) selecting the right technology, and (3) effectively managing computer people. Network computing increases the management difficulty because it mandates close cooperation between the different computer disciplines. For example, when things were simpler, application designers could ignore the specialists who handled the esoteric technical details that kept things humming. Now, performance depends as much on the network architecture and infrastructure as on the application design. This network computing characteristic mandates a close working relationship between the applications and technical staffs. Teamwork has become the order of the day.

Many computer professionals fixate on a single solution set. When given a chance, they love to dazzle people with the latest technology in their specialty. They tout this new advance, promising that it speeds development, reduces cost, and provides a quantum leap in functionality. Too often, without getting a second opinion, management begrudgingly approves a large project using the technology and then moves on to other concerns, ignoring the project until it's too late. They are managing on faith instead of controlling events because the mystique of computers causes many to neglect good management practices.

When a new computer doesn't deliver on its promises, someone has not been doing their job, but management is not sure who. The usual suspects include the seasoned professionals, derisively known as dinosaurs, and the young whippersnappers who enthusiastically push new technology. Neither of these groups deserve full blame. Their reluctance to work together, however, does magnify the difficulty of getting new applications out to the workplace. So much quarreling goes on within the computer community, it's a wonder anything gets accomplished. To quicken the pace and move with surefooted ease, it is necessary to get the various computer factions working as a cohesive team.

The AlliedSignal culture firmly believes in root cause analysis. Quick fixes, expressions of remorse, or promises to do better next time are not enough. Management wants to know why things went wrong and what needs to be done to preclude a recurrence of the problem. AlliedSignal's first foray into network computing, mentioned in the beginning of this book, was less than a sterling success. A postmortem quickly diagnosed one of the major problems to be overreliance on a few individuals with a narrow perspective on computer technology.

The Big Fix

The fix didn't come easy. To address the root cause, the necessary actions included a major restructuring of the computing technology function. Many people knew about the brewing fiasco, but warning signals never reached business leaders. A check-and-balance system needed to be put in place to preclude technical bias and provide executives with an early warning system. The company decided to leverage an existing program. AlliedSignal had already consolidated their mainframe data centers to reduce cost. The program provided another, unanticipated benefit. It created a core of computer expertise in a central location.

The answer became obvious; Extend the concept to include all computer and network disciplines. The Computing Technology Center (CTC) changed from a mainframe computing center to a "center of excellence" for every computing and networking technology. The CTC recruited the best experts in every discipline for an expanded mission to provide corporatewide guidance on all technology deployment decisions.

Why did AlliedSignal land on this particular solution? Because the company exposed one of the root causes to be fighting within the computer community. The revelation came from comparing a bungled project with a successful network computing application. The successful application avoided brand-X middleware and arcane tools, provided network infrastructure, had a realistic architecture, and met the business process requirements. Modifications in the architecture eliminated design aspects that would have impaired reliability or responsiveness. Further investigation reveled a team-oriented project leader who consulted everyone in the free world. She consistently reviewed decisions with other technicians and end users.

The project leader for the poorly managed application stiff-armed everyone. He moved with assurance because he was confident he had superior knowledge. To prove his brilliance, he eschewed market leading products and searched out boutique software from small shops. He constantly disparaged other people's skills and accepted advice from no one. Despite his project's dismal progress, he initiated technical debates with anyone who questioned his approach. Like many of his ilk, he loved to fight endlessly over esoteric details and technical trivia.

The computer technology discipline doesn't need any more fighting. The time has come for an armistice, but it will take more than shoving people with different skills into the same cubicle in the hope they become soul mates. If you have concurrent engineering initiatives, then you already know how challenging it is to get engineers and manufacturing people to work together. It won't be any easier with your computer professionals.

The Raging Culture War Inside the Information Revolution

The energy and drama of the information revolution conceals a serious cultural conflict. The three cultures include the mainframers, the personal computer technicians (PCers), and scientific computing. These cliques prevent teamwork by an emotional aversion to recognizing the skills and technical knowledge of others. Each group defends fiefdoms and enclaves with tactics that would make a seasoned ward boss proud. In many situations, political correctness replaces reason, bias taints decisions, and the ordinary elevates to the ethereal. No wonder managing computer people seems so difficult.

To be a successful manager, you need some background on how the culture war got started. First, the conflict has been seething for over a decade. Although it is not a shooting war, this conflict does include the usual cast of characters: combatants, allies, partisans, collaborators, spies, and cheering crowds on the sidelines exhorting the participants to fight until victorious. The battle ranges across familiar territory: the haves and have-nots, generational disputes, money and power, intellectual leadership, and the government's role in all of this.

The factions battle endlessly over ideas and ideals: ideas on how to use computers and the ideals within which a society should function in a technological era. This power struggle is for nothing less than the leadership that will take us into the twenty-first century. Can computers really cause this much ruckus? Of course not, computers are merely inert boxes without a clue. However, the people who use them sometimes can engage in some real unpleasantness.

As in any struggle over culture, the direct combatants comprise a very small number of zealots, but the repercussions spill over to swamp us all. This ugly war, like every war, produces a few socially redeeming values such as rapidly advancing technology, the weeding out of weak players, and the induction of bright, young recruits to carry on the struggle and convert the masses. But like every revolution, this one has gone to excess and become self-destructive. You need to negotiate the peace. Why you? Because your money fuels this foolish conflict.

The animosity between the different computer groups increases the difficulty of introducing new information systems. No major supplier provides architectural leadership, technology advances at an unprecedented pace, and the computer technicians have broken into armed camps intent on gaining leadership of the information revolution. Management can handle the first two issues after they control the third by putting a stop to the bickering. These people run amuck because we, as managers, haven't figured out how to control their work. Once you understand the cultures of your computer people, you can easily manage their performance. You need to grasp the underlying biases of the proponents waging these never-ending battles over how we use computers. The various antagonists constantly throw up flack and chaff to obfuscate your ability to hold this discipline accountable.

The technological differences mean less than the perspectives, biases, experience, and problem-solving techniques of the proponents of each discipline. Each has its own agenda, dedicated trade press, gurus, functional mentors within the organization, industrial allies, collaborators setting industry standards, and gatherings in the form of seminars and conferences designed to refresh proponents' fervor for the cause. Although the followers of each of these cultures believe that they own a solution set with universal applicability, it's simply not true.

A Cold War Turns Hot

The formal declaration of war occurred on January 22, 1984. During half-time of Super Bowl XVIII, Apple ran a commercial to introduce the Macintosh personal computer. The commercial ran only once on national television. This expensive production showed legions of identical workers, all dressed in gray concentration camp clothing. With cleanly shaven heads and darkened eyes, they stared uncomprehensively at a giant television broadcasting a dispassionate speech delivered by Big Brother.

Into this gray-toned, Orwellian world charged a pretty, young woman dressed in bright red track shorts. She effortlessly carried a large sledgehammer. Quickly sprinting through the benumbed zombies, she flung her sledgehammer at the huge television screen and shattered it to smithereens. The tag line for this powerful imagery was "Macintosh, for the rest of us." This was nothing less than a broadside across the bow of IBM's flagship PC and the IBM culture. The IBM personal computer had taken the market by storm and set the standard for desktop computers, just as IBM had set the standard for large-scale computing. The commercial sent a clear message that a new breed was challenging IBM's stale and unimaginative leadership.

The commercial for the Apple Macintosh exemplified marketing acumen. It built on the cultural chasm that already existed between the traditional big computer proponents and the kids who grew up with the Apple II. The perspectives of the two groups were, and still are, completely different. The traditional computing professionals viewed their job as providing a business tool in a structured organizational environment. The PCers saw their machine as a social phenomenon that unleashed individual creativity and evened the odds by providing computer power to everyone.

At the dawn of this revolution, young hobbyists, not computer professionals, pushed the embryonic personal computer beyond the state-of-the-art. Many of these hobbyists proudly claimed the title of hacker. They constantly pushed the envelope of their equipment—often right past big computer security systems. Through use of a telephone and a few wily techniques, hackers energetically broke into every computer they could find. They reveled in outwitting the professionals.

Hackers harbored a visceral disdain for mainframes and all the rituals surrounding them. They objected to the rules and bureaucracy that denied access to these machines. They loved their unrestricted access to personal computers and believed these marvels of technology should be available to everyone. They saw the PC as a liberating social force that would democratize the power that came from free and open access to information. The cant said that personal computers would change the world. They were right.

A Change for the Better

Computers, including personal computers, are changing the world. So did the internal combustion engine. They both opened up the amount of space in which we exist; the automobile increased our physical space and the computer opened our intellectual horizons. Undeniably, each brought social change, but revolutionize our entire existence? Hardly. These inventions merely represent technological tools that leverage our innate ability to move and think. The early models of both were difficult to use and variations came along later to meet every purpose. Each increases our society's ability to produce goods and rewards us with new consumer products that we quickly take for granted.

The major difference between the internal combustion engine and computers is the emergence of fervent subcultures aligned with derivations of the basic technology. True, automotive enthusiasts include racers, street rodders, truckers, and people who restore antique cars, interspersed with those who take loving care of the family sedan; but these divergent enthusiasts do not believe their passion will change society. The difference could be the maturity of automobile technology. Possibly, but I believe other social phenomena account for the stridency with which computer subgroups fight to perpetuate their narrow perspectives. To achieve the systems we need, we must understand the motivations of the protagonists.

Let's look at some of the personalities that are probably clamoring for your attention and largesse. Keep in mind that I am using stereotypical examples because it makes it easier to explain the emotions that stoke this conflict. We'll begin with the mainframers. They used to be the vanguard of the revolution, but now they are on the defensive after losing numerous battles with more aggressive rebels.

Dinosaurs in Need of Pest Control

Mainframers came to the party early and didn't realize they had joined a crusade. They just saw a new machine and fell in love. Introduced in the mid-1960s, the IBM System 360 claimed their affection. This expensive piece of hardware needed extensive air conditioning, isolated power, and a special raised floor to hide all the cables that lashed the pieces together. Companies displayed pride in their fancy, new computers by putting them behind glass windows so people could gaze at these marvels, without the risk of anyone actually touching one. Thus was born the "glass house"—look, but do not touch. A high priesthood developed around the glass house, and only they could enter the inner sanctum to apply the sacraments that would induce the machine-god to spew forth wisdom by way of reams of paper.

The Plebeian Elite

Who was this priesthood? A bunch of bright, but ordinary, people who became enamored with this new technology. Computers existed prior to the System 360—and electronic accounting machines (EAM) prior to that—but the 360 design included a new architecture that made computer solutions possible for many more applications. The young and smart raced to get into the field because it was exciting, it paid well, and no established hierarchy blocked their path to advancement. If they had bothered to look back, they would have seen that they were displacing a prior generation of technocrats who had thought electronic accounting machines held the key to a great career. Some of the EAMers made the transition, but for the most part, mainframers were leading-edge baby boomers—firstborn children of the returning heroes of World War II.

These early technicians went through school when colleges started offering a few computer courses and the excitement over the System 360 peaked. Despite the lack of professional stature, young people perceived computers as a glamorous career. They ruthlessly competed for the available jobs. Typically, mainframers, as the oldest in their families, have now gravitated to senior management positions in computing technology. This profile is completely different from that of their younger brothers and sisters who came to the party late and joined the information revolution through the back door of personal computers.

Freight Train Applications

Like their equipment, mainframers are into heavy lifting. For years they have been doing high volume, mission-critical applications that must be up for their businesses to do work. When the mainframe, or an application, goes down, these technicians get angry calls from the high and mighty. Executives scream that they can't ship product, they can't build product, they are losing revenue because they can't make airline reservations, security or commodity trading has stopped, customers can't get cash from ATMs, or retailers can't check the validity of a tendered credit card. The mainframers' creed is reliability and maintainability. They build systems to last. They can be obsessive about testing. When it breaks, they want any number of people to be able to fix it in a flash. After they complete a system, they want to hire a neophyte off the street to maintain it.

Their technology works. It's just that it's expensive, hard to change, and not user friendly. Personal computer software changed the level of expectation. This, more than any other reason, explains why mainframes get so much bad press. Mainframers have a hard time understanding this; they believe in functionality, high volumes, quick response time, reliability, maintainability, and consistent, predictable service levels.

Now everyone wants a graphical user interface. They know their systems are homely, but beauty is only skin deep. And in any case, they've found a solution; they'll shed a few pounds, buy some new clothes, and use a little plastic surgery to improve their outward appearance. In other words, they've started putting GUI front ends on their old mainframe systems, hoping nobody looks behind the curtain.

The powers-that-be want more than a pretty new face: they want systems built in months, not years. They've been led to believe that the mainframe is dead and you can easily throw something up using new technology. A caveat exists, of course. You must have the right kind of people to build network computing applications. Every book and conference speaker says that 40 percent of the current programming staff needs replacing. Mainframers just can't hack it anymore.

Bright, young, personal computer technicians need to be recruited and nurtured to build the brave new world. The unanimity of opinion and stark prejudice makes the traditional applications people skittish. Much of the criticism hits too close to home, so mainframers have adopted a siege mentality. They are slow and expensive. They do spend too much time keeping the old stuff from breaking. They want to work on new things; they just can't get relief from fixing this and repairing that. The hardest part for them to accept is the emotional discarding of everything they know and all they have ever done.

The Whippersnappers Want an Elephant Gun

The personal computer professionals can't figure out why mainframers still consume so much of the precious computer budget. Everything they propose is in seven figures. They really can't comprehend the mainframers' obsession with six-phase development cycles. Don't they understand the new world embraces rapid applications development, prototyping, and disposable code? You have to move fast to keep up with the accelerating business world.

There's no room, nor patience left, for their clumsy development processes and testing procedures. And what about that documentation? If they knew how to design good systems, intuitively obvious ones that use a graphical user interface, they wouldn't need so much documentation. The biggest problem with mainframers is that they work with the wrong people. Instead of dueling in the basement, they need to come upstairs to discover what the people running the company really need. Oh well, they won't be around that much longer anyway.

There Goes the Neighborhood

Personal computer technicians tend to be a rebellious lot. Personal computers attracted them because these machines challenged the existing norms. The freewheeling atmosphere and excitement appealed to their nature. The Apple II, or less likely, an early IBM model, claimed their affection. You could touch it, you could play with it, and you could even change the way it worked to match your mood. You didn't need to ask anyone's permission to use it and no one looked over your shoulder. The personal computer provided the perfect foil for people reared to challenge authority and question everything. Management couldn't ask for a better set of attitudes to drive change and innovation.

Small Is Beautiful

The Apple II computer created a large cult of young enthusiasts who became enamored with their new machine. By today's standards it was not much, hardly more than a large paperweight. However, back in the 1970s, when the uninitiated saw it for the first time, they immediately recognized it as magical. The Apple II didn't require heavy financing, it talked to their friends' computers, and it played games. Games originally attracted the young, but they rapidly graduated to simple programming. If games and simple programs encompassed the entire world of hobbyist computing, then these budding hackers would have stagnated behind the doors of their gadget-laden bedrooms.

There was more, a lot more. Computer clubs delivered on the Apple II's promise of a self-contained, batteries included, lifestyle. These computer clubs acted as the catalyst that transformed thousands of bashful, young techno-groupies into a highly trained army of computer experts. Just like boot camp, these clubs instilled a common set of beliefs into their recruits. The members gathered frequently to trade tips and ideas, share industry gossip, and pass around pieces of esoteric code that did wondrous things.

As the PCers reached adulthood, they came into organizations that were adopting the desktop wonders and increasingly needed to network them together. Since their first day on the job, PCers have been working with people who could justify expensive equipment on their desks; people who did more than repetitive, mundane tasks. Their solution set evolved to meet the needs of this customer base. As personal computers migrated from the staff members' cubicles to executive offices, the PCers began to work with upper management. Executives have little patience for reading thick manuals, so PCers solved the problem with ease of use.

Even before graphical user interfaces, personal computer software made the mainframe variety look clumsy and unsophisticated. Now, with Windows and Macintoshes, the comparison is ludicrous. The broad acceptance of graphical user interfaces instilled them with confidence as they started to take their solution set out to the rest of the world. Now, they're starting to discover some serious holes in their technology, but they've been conditioned to expect fast, innovative solutions. They're right, things do move at lightening speed in the microprocessor industry.

The Academics and Their Machines

Anyone can be smitten, even an engineer or a scientist. They fell in love with the Digital Equipment Corporation's (DEC) VAX family of processors. Actually, infatuation better described their liaison. Engineers are a fickle lot, quickly changing their allegiances whenever someone introduces a faster machine.

DEC sold "roll your own" computers. A phone book would lose a heft test with the manual that listed all the available features and options. Nobody challenged what engineers ordered because they didn't understand it: not the mainframers, the purchasing department, accounting, or even their own bosses. What did all those features do?

Crunch numbers—lots of them. The VAX could handle really big numbers, or exceedingly small numbers. The VAX didn't sport a blue uniform, but it was still faster than a speeding bullet. At the time, the IBM boxes ran a poor second when they competed against DEC equipment.

Scientific computer users display little brand loyalty; they buy the fastest box on the market at any point in time. Don't expect them to take a salesperson's word for the relative speed of a computer. Someone who wants to sell to this market needs to be ready for painstaking tests that benchmark their box against everything else. Any engineers or scientists worth having on your payroll think their work is the most important thing around. When they want to get to the computer, they want it now and they don't want to share. Finally, they have an insatiable demand for processing power. No matter how quick a machine you buy, they can consume it faster than the depreciation schedule.

Engineers and scientists view the computer as an adjunct to doing their job, merely a tool to extend their intelligence. No matter how much time they spend on computers, they do not think of themselves as computer professionals. This doesn't mean they don't have extensive computer knowledge, only that they do not want to be classified as computer technicians. In fact, engineers have quite a bit of experience in tying together different brands of computers. Another common characteristic is their desire to isolate themselves from all other forms of computing. They fear that if someone requires them to integrate their computing capabilities with business-oriented computing, then they will be less free to pick and choose from the latest and greatest. This is not an unfounded fear. The integration of all the computing capabilities within an organization requires enforcing certain standards, and standards limit choice.

The Hatfields and McCoys

The technical computing crowd wants to be free to buy the fastest box on the market without having to take other people's needs into account. The PCers inherently fear being hemmed in by rules. This makes either party reluctant to work with the mainframers who constantly harp about standards, security, and controls. This long-running feud hampers your ability to build a digital organization. This cultural war has progressively escalated and can cause peripheral damage throughout your organization. It's not fun being hit by friendly fire. Getting these people to quit quarreling and start working together on the same team won't be easy, but it's a worthy objective.

Mainframers are jealous of the PCers' easy access to the executive offices, their snazzy user interfaces, and the quickness with which they can dispense a never-ending stream of shrink-wrapped applications. The PCers have grown tired of users demanding that their systems be as reliable as antiquated mainframes. The engineers and scientists can't believe how often they have to fight off encroachments onto their turf. This scuffle is over more than technology. Like any political war, the zealots fight to control the debate. Despite the incantations by the extremists, computers do not represent a social phenomenon. They are merely machines.

Am I exaggerating? Of course. In more and more companies, these three groups are starting to work together and share their perspectives and technical approaches, but we have a long way to go. By we, I'm talking about managers responsible for directing our organizations to be competitive, smart, and responsive to business needs. The last thing we should do is choose up sides. We need to induce—force if necessary— these disparate groups of professionals to share their experiences and work together toward practical solutions. Why? Because they all have a lot to offer.

Peace in Our Time

Today, most mainframers have personal computers sitting in front of them and PCers develop transactional systems. Events will require scientists and engineers to work with both groups. The fighting still goes on, however, and we need to settle things down without losing the vitality and vigor for change these respective professionals bring to our organizations. Increasingly, managers recognize that mainframers and

PCers control much of the activity that needs integration with engineering computing to meet the speed-to-market imperative.

Lead from the Top

How does AlliedSignal address this problem? Aggressively. Exhibit 9.1 shows the AlliedSignal approach. It starts with executive leadership. Larry Bossidy insists on a "one company" culture. Collaboration and teaming get encouraged with handsome rewards and recognition. He requires company councils to be cross-functional. The Executive Information Systems Council includes engineering, and the information systems function has membership on every business council. The evaluation of computer professionals happens twice; once within the computer technology community and a second time by business leaders. The top-down leadership builds a culture of cross-functional cooperation that permeates the company.

The Computing Technology Center represents another element of AlliedSignal's approach. By concentrating a central core of expertise that spans the entire gamut of computing, AlliedSignal forces communication and collaboration between the factions. To avoid an "us versus them" atmosphere, the CTC must always act professionally, represent the company's interest, and display a team spirit with the business unit organizations.

Exhibit 9.1
Team Building

Executive leadership.
Co-locate different technical skills.
Cross-training.
Fact-based process.
Flexible teams.
Broad technical awareness by management.

Spread the Wealth

Collaboration between the computing disciplines requires common frames of reference. AlliedSignal invests heavily in cross-training and education. Every computer technician does not need to be an expert in every specialization, but a general grasp of the fundamentals lubricates communication with colleagues. The CTC staff attend formal training courses outside their specialty on a fairly regular basis. Participation on cross-functional teams and temporary assignments to other departments spread knowledge by "on-the-job" training. Technicians give descriptive briefings on new technology to spread knowledge and hone presentation skills. At a higher level, the Information Systems Human Resource Council develops and moves people throughout the company. This cross-pollination encourages teamwork, provides common frames of reference, and disperses expertise.

Additionally, the CTC career development program includes a procedure wherein technicians can earn a company-paid course or seminar in a subject of their choice. If they voluntarily join a team outside their discipline and put in a specified number of uncompensated hours, the company pays for off-site training. This popular program sets a tone that encourages multidiscipline skills and puts a share of responsibility for career development on the employee.

As a Matter of Fact

Computer people are analytical. This characteristic saves the day. AlliedSignal designed a technical evaluation procedure. Another AlliedSignal tenet is "act on fact." Whenever a technical debate occurs, the proponents must collectively produce a technical evaluation document. This forces the argument away from emotion and opinion to fact-based analysis. This process works surprisingly well. For example, when a network computing application has performance problems, instead of finger-pointing, a cross-functional team must do a thorough analysis and publish the results. The technical evaluation process includes a "get well" plan that lists the specific actions necessary to rectify the situation. Blame is not assigned. Instead, a section of the report calls for "lessons learned" so the problem can be avoided in the future.

Team Play

Cross-functional and cross-discipline teams provide the best protection against computer project failures. Network computing is complicated. No one, individually, has the experience to design a modern computer system without assistance from other experts. Don't let one individual lead you astray. Collectively, a group of sharp computer professionals will always design superior systems.

Early in my career, I was assigned as a dedicated programmer to the director of production control. We kept trying to design a closed-loop system for part shortages. Every time we thought we had closed the last hole, the shop floor found a new way to beat the system. One day, in frustration, the director said, "Why can't we figure out a way to keep them honest?"

"We're a couple of smart guys," I responded, "but collectively, four thousand people are smarter than us."

Remember, two heads are better than one, and six or seven really sharp people can spot almost every booby trap in a design.

Sharing Knowledge

Success in deploying sound business technology requires leadership and the astute management of computer professionals. If you directly manage computer professionals, you need at least cursory awareness of different computer technologies and their application in the world of work. I know it's hard to keep up. But if you want to avoid inadvertently going down the wrong path, invest a good part of your time keeping current.

I previously mentioned that CTC technologists frequently give awareness briefings on their specialty. These were originally designed to update technical management. Only later did we realize the benefits of educating everyone. This technique has numerous advantages over vendor briefings and outside seminars. Your staff gets presentation experience, they learn to simplify complex subjects, your own people present caveats that vendors never bring up, and you learn what your own people think about new technology. These presentations also provide an opportunity for you to direct resources without having people spend untold hours on nonstrategic issues.

Putting this concept in place only requires that you schedule regular briefings that rotate between groups. Give clear directions that the presentation must be polished and substantive, but at an "awareness" level (the same type of presentations you make for your bosses). Don't worry if the first ones don't go too well, they'll automatically get better if you keep at it. Make sure the presenters understand that these briefs help prepare them for career moves into management. Start with an area of intense interest and select someone who knows the technology and has decent presentation skills. You'll be surprised how much knowledge resides within your own organization. Besides, once the staff knows your expectations, they'll be poring over trade journals, studying technical manuals, and attending vendor presentations. Concentrate your time on events that allow people to mix with peers and business leaders to prevent myopia.

It's time to move on. Let's stop talking about computers, and start discussing what to do with them. To move effectively, you need to get your "bits and bytes" people working with your applications staff. When computer specialists start working together, instead of fighting, the surprising synthesis accelerates productivity faster than anything else imaginable—make sure this happens in your company first.

FURNISHING YOUR DIGITAL ABODE

Manage Your Application Development

One morning I shot an elephant in my pajamas.
How he got in my pajamas, I'll never know.

—GROUCHO MARX

A digital organization needs more than the ability to compute and move ones and zeros between two points on the compass—it needs applications. Computer programs must tell all this expensive hardware and infrastructure what to do and when to do it. You can buy the programs, you can make the programs, or if you're really smart, you can assemble them from standard components. Whichever approach you use, you still need to manage the development process so your digital organization comes fully equipped with applications that leverage your business strategy.

Implementing applications faster, and for less money, tops most executive wish lists. The past record for application projects may shake your confidence, but applying a little common sense, disciplining the process, and providing leadership makes the future much brighter. The elusive challenge is to develop a process that repetitively leads to success.

179

Why Applications Are Difficult

Laying down hardware is relatively simple. Even provisioning that hardware with system software, tools, and utilities doesn't present a big problem, especially once the software houses bring to market the products we need. (System software lags hardware for the same reasons you can't get applications out to the workplace as fast as you would like. People in the software business haven't found a secret formula, nor has anyone else broken the code, for quickly producing software inexpensively. The tools keep getting better, but we raise the bar often enough that we continue to be disappointed.)

The ease of installing hardware and shrink-wrapped software sets a level of expectation impossible to achieve for business applications. You need to pick the right target of opportunity, acquire or build the application, integrate it with your other applications, and then implement it. Implementation presents the bugaboo. Training, process changes, integration issues, unexpected design omissions, and initially slow performance make implementation troublesome and at times fatal.

Applications moving out of development into the workplace almost always encounter scaling problems. In the absence of inherent design flaws, some frenzied tuning, quick fixes, and hardware upgrades usually get performance to an acceptable level. But first impressions linger, and many people who resist the process or organization changes eagerly pass along the complaint that the new system doesn't work. These shakeout jitters don't cause the stillbirth of new applications—implementation difficulties occur because of changes in the business processes.

If an application does not change the organization and its processes, then you probably did an incomplete job. When you do change the business processes, you change people's behavior. You need to train them in the system and the process changes, but more important, people must readily perceive the new processes as an improvement. Technical issues don't normally kill application initiatives, poor design does the dirty deed. A digital organization requires a solid supporting infrastructure, but it must also effectively manage applications design from inception to full integration of the process into the organization's daily routine.

Crux Deluxe

The base technology of your applications does not present the biggest hindrance to progress. We can solve the technical interface problems and data incompatibilities. Data and interface issues become less of a problem every day as new products continue to fill the market need to exchange data between different types of computers. Simplifying your infrastructure and standardizing your interfaces makes this problem manageable. Program logic represents the real obstacle.

Computer programs dictate your business processes. When you want to change the process, you must change the programs. Since old programs don't adjust easily to a new lifestyle, you authorize endeavors to implement new systems. But it's not that simple. You can't replace everything simultaneously and these old programs have deep interdependencies that botch attempts to separate the good, the bad, and the ugly.

How do you solve the logic interdependencies and data integration issues? I don't have a simple answer. Here's why. Corporate leaders frequently complain about the lack of integration in old systems, but these applications are actually integrated to an extreme. The difference is that traditional systems were integrated within a business unit, while many new applications attempt to integrate the same function across divisions and also vertically to be consistent with the corporate reporting structure. Integration issues present a problem because the business model changed. AlliedSignal's Business Services invented a new business structure that separates services shared by many business units. As a result, the company needed to untangle the programs that assumed a different model. Business model changes, not technology, cause the greatest difficulties with application development.

Managing the Process

Building new systems while maintaining order and containing cost has proven to be a hit-or-miss proposition for most organizations. Applications development challenges the most astute manager because it combines the technical and analytical with organization behavioral issues.

Sophisticated computer technology amplifies the issues of process change. This characteristic makes applications a general management responsibility, not something relinquished solely to computer professionals.

Application customers complain that computer technicians don't understand the business, but application development leaders feel shunned by business managers who can't make the time to provide proper guidance and leadership for major projects. It's scary to feel abandoned when you're responsible for designing major changes in an organization. I know, I've been there.

Improving the development process requires business leaders to take a more participatory role in the development of applications. A digital organization's objective is to leverage organizational design, world-class processes, and people's skills with innovative computer systems. This requires a total team effort between computer professionals and business leaders. In this chapter, I'll cover the basics of applications development, with a focus on how business leaders can improve the process. Exhibit 10.1 lists the major areas to be discussed.

Project Management

This story is about my first serious design effort. After I received a promotion to senior analyst in the early 1970s, my boss called me in to tell me that my next assignment would be working with an analyst named Jack. He didn't know how Jack did it, but he knew he designed good systems and users loved him. Jack was about to start a new project for an

---

Exhibit 10.1
Managing Applications

Project management.

Propagating common systems.

Development methodology.

Cheaper, faster.

Data versus process.

---

online storeroom control system, and I was to shadow him and learn his methodology.

Jack's methods were less than orthodox. The first thing we did was set up appointments with every manager that had the slightest interest in the project. Jack always gave me specific instructions on what to wear for each meeting. If we were meeting with accountant types or the financial executive, he told me to wear a starched white shirt and conservative tie. When we met with manufacturing or production control, we loosened our ties and rolled up our sleeves. In Jack's mind, the way we dressed didn't represent pandering, but a display of empathy with the customer's culture. He didn't want superficial differences to distract from an open discourse.

At each interview, Jack dutifully asked each manager what he or she wanted in the new system. He conspicuously took extensive notes on everything they shared with us. After we finally made it through this management gauntlet, Jack took me aside and made a big show of shoving his copious notes into a drawer.

"What a minute," I said, "what are you doing? Don't we need to review those notes to design the system? That's our entire record of what everybody said."

"Not yet," he said. "Now we're going to start this project for real. I want you to show up tomorrow morning at 7:00 A.M. and wear casual clothes and steel-tip shoes. We're going to spend the next month in the storeroom."

And we did. We got ourselves a couple of desks in the stock crib and designed the system on the shop floor. We spent many hours actually doing the work of crib attendants and questioned everyone on every detail of the job. As we gained an understanding of the work flow, we redesigned the processes to improve efficiency and wrote the specifications for the system. After we finished our research in the factory, we returned to our offices to formalize the design. Jack now pulled the meeting notes out of his drawer.

I asked Jack, "How are we going to handle this? We already designed most of the system and we haven't incorporated the requests from management."

"Don't worry," he answered. "The requirements from the factory won't be in conflict with what management wants. And we haven't designed most of the system, we've only nailed down the operating procedures for the transactional activity."

Again, Jack was right. The managers had asked for controls, metrics, activity summaries, exception reporting, and process cost reduction. Everything they asked for required accurate and timely transactional records as a prerequisite. We had already defined the transactional part of the system, so our next step was including the necessary features that would give management visibility of the processes, trends, exceptions, and cost.

Jack next scheduled another round of manager meetings. At each meeting, Jack presented our design. Throughout the presentation he interspersed comments like, "As requested during our interview process, we've included the following feature." The managers didn't remember whether they had suggested the particular feature or we had picked it up in one of our other interviews, but the comments showed that we had listened and responded to customer requirements.

After the second series of meetings, Jack shared with me his philosophy of applications design. "Listen," he said, "applications development is a management job, not a technical task. You manage the requirements to realistic levels; you manage the level of expectation and ownership; you manage the design process to get a system that works in the real world; you manage the programmers to make sure they stay on script; and you manage test, training and implementation to make sure the system meets schedule and user expectations.

"Never design a system to only one class of customer requirements. Talk to everyone. Managers are paid to provide organizational direction, resolve problems, handle exceptions, and improve business performance. They give important guidance, but don't forget to also interview the people who do the actual process. Spend some time in their shoes, and figure out what can be changed to increase productivity. Concentrate your initial design on the repetitive processes. If you get that wrong, everything else will flounder.

"One more thing; always pretend the person you're talking to has complete say in how the system is designed. Don't bring up other people's requirements. To get user buy-in, you need them to have a strong proprietary interest in the application. The crib attendants don't appreciate all of management's concerns, and managers just assume we'll do a professional job on the routine. If a serious requirements conflict does arise, settle it in a private meeting with only the people involved."

Jack taught me that management meant managing events to a desired outcome. That means managing everything that influences the course of events. He adjusted his style and approach depending on what would work in a particular circumstance.

Technology projects fail because of mismanagement. However, a failed project may not be entirely the project leader's fault. The breadth of impact of modern computer systems requires a team effort. No one has wide-enough experience to handle all the complexities of a sophisticated systems project. A mix of skills is necessary.

Today, most application designers have very little business exposure, yet they design processes that significantly impact the organization. How did we get into a situation where computer analysts, with insufficient business process knowledge, determine how to run our businesses? It started in the 1970s when the growing complexity of computer technology changed the perspective of design analysts from business to computer science. Jack started his career in production control, not migrating over to computers until after he learned how the company operated. The once common practice of recruiting bright professionals from other fields died when the technology became so complex it warranted full-time study.

During stints as a curriculum adviser to two universities, I recommended requiring English, speech, and business courses for a computer degree. I argued that computer professionals need communication and application skills to sell their wares and move into management. Although I made friends in the liberal arts department, I never convinced the powers-that-be to broaden the curriculum with more nontechnical courses.

Today, many project managers have not worked outside the computer field, nor can you be certain their formal education has given them a foundation in basic business principles. Despite acquiring a college degree in computers, attending seminars, and devouring every trade journal in sight, it's difficult for today's computer professionals to keep up with technology. It's unrealistic to expect them to be an expert in accounts receivable, production scheduling, engineering design, opening a bank account, and preparing a sales call.

Worse, we promote the best technicians into management without providing even rudimentary training in basic management skills. Many times, we end up frustrating excellent technicians by asking them to perform a new managerial function without proper training or guidance. How careful are you about assigning someone to their first department manager role? How much training, mentoring, and counseling do you provide for high-potentials in other fields? Do you use the same care cultivating project leaders responsible for computer systems with competitive ramifications well beyond the development cost? I hope you answered yes to the final question. Your highest priority should be searching out good project leaders and developing their management skills, because they hold your career in their hands. Good candidates are easy to recognize, they have a string of successes behind them and the scars to prove it.

My recommendation is to restart the practice of interdiscipline transfers, but even back in the early days, it took several years to train and educate a professional from another discipline on computer technology. If you start moving people, it will take a long time before you see a benefit. Nevertheless, start a formal program of transferring people between your computer and business functions.

The trend in applications is toward purchased software and reusable code. As the trend develops, less technical knowledge will be needed by system designers, allowing organizations to again select project managers with a primary focus on business processes. If you start now with a program to systematically move people between the computer functions and business areas, then you'll be ready for the day when it's possible to assemble applications from standard components. In the meantime, assign competent and motivated businesspeople to system project teams to augment the technical staff. The additional investment will prove beneficial to the entire organization.

Common Wisdom

Common systems provide the fastest way to achieve a digital organization. An enterprise project may take longer than a point solution, but the entire organization elevates to a higher plane of performance. Enterprise applications need three prerequisites: (1) a solid, standardized infrastructure, (2) a mature and competent applications development organization, and (3) general management responsibility consistent with the level of commonality. Omitting the third prerequisite results in elongated, and frequently doomed, development efforts. You can't satisfy everyone, so someone needs to act as the enterprise functional leader. AlliedSignal accomplishes this either with line responsibility, as in Business Services, or with a matrix structure. AlliedSignal's corporate office includes vice presidents for each business function and they assume the responsibility for making the hard compromises necessary to get common applications out of the requirements phase.

Don't go overboard on enterprise applications. Make sure the process can be common and your technology base supports the design. Temper your zeal for taking a good idea too far by keeping in mind that islands of automation exist because they're doable. Point solutions give business relief quicker, provide a test bed for new technology, and iron out process logic for follow-on enterprise solutions. Arbitrarily dismissing islands of automation is a mistake; sometimes they can be your best friend. You can't tackle everything at once, so allowing occasional point solutions buys time until you reach that particular process. Do set corporate guidelines that specify when point solutions may be used, and when they're forbidden.

Pollyanna Will Not Come to the Rescue

Many companies search their own enterprise for the "best practices" and "best-in-class" systems. As you find good implementations of best practices, don't automatically assume you can propagate the software that supports the process. The system may work fine when confined to its own environment, but develop a terminal illness when transported out of its natural element. This next law may be painful:

Law #11   Serendipity does not apply to computer systems.

Sorry, but this one seems cast in concrete. Worse, there are a number of corollary statutes:

- If a system was not designed from inception to be portable, it is not.
- If a system was not designed from inception to be scalable, it is not.
- If a system was not designed from inception to be intuitive, it is not.
- If a system was not designed from inception to be maintainable, it is not.
- If a system was not designed from inception to be integrated, it can not.

Law #11 means that computer systems must be purposefully designed for every characteristic you desire because it will not happen by accident. Although serendipity does not apply to found treasure, that doesn't mean it has no value. Keep in mind, flawed process design mortally wounds applications. Your discovered application has already ironed out the process logic, which can save you time and money, as well as helping you avoid design errors. The application probably provides an excellent template to copy. Even better, if you've established sound development principles throughout the organization, much of the foundation may be reusable.

Whether you buy or centrally develop enterprise applications, making the system an integral part of all your businesses goes faster and smoother if you assign a permanent implementation team. This team should comprise dedicated people, separate from the design and development staffs. Implementation of complex business applications has a steep learning curve. Once a team has completed an installation, they become much better the second time. One caution, making systems implementation a repetitive task takes a heavy toll on the installation team, especially if—like AlliedSignal—your business units are spread all over the map. The work is fast-paced and stressful, and it doesn't help that they are away from home months on end. Fatigue became a major factor on AlliedSignal's roving team of implementers. My suggestion is to schedule rest breaks in the implementation cycle, or overstaff the team to allow rotations home on a periodic basis.

Development Methodology

I initiated a study once to compare the history of several application projects. The results uniformly followed the pattern shown in Exhibit 10.2. This pattern held, whether we purchased the software or developed it in-house. Cost was not a major factor at the beginning as a few senior analysts determined the requirements, sought project approval, and did the external design. Elapsed time became the driver during this stage. During development, we either forked over the money for purchased software, or put an army of programmers on the project. If the designers handed over a decent plan, then development occurred relatively quickly,

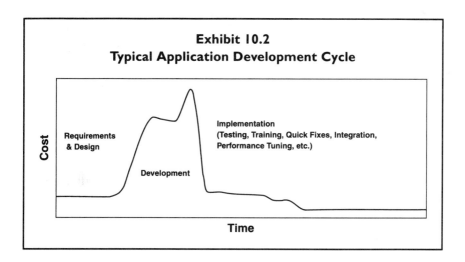

Exhibit 10.2
Typical Application Development Cycle

Cost

Requirements & Design

Development

Implementation
(Testing, Training, Quick Fixes, Integration, Performance Tuning, etc.)

Time

but cost went through the roof, especially at the end, when additional programmers and analysts piled on to make schedule.

The study's startling revelation was that the implementation phase seemed immortal—we called it the "Eternal State." After the majority of the development team went on to other tasks, a few key people lingered, struggling to fully implement the application. Cost continued to build, but the real problem was bringing closure to the project. Doomed projects seem to die a slow, painful death in this stage. If a project fails during implementation, the money has already been spent. You not only lose the application, but incur a high opportunity-cost that puts you further behind the competition. The Eternal State slowly eats away at management's patience and destroys user confidence.

We conducted this study to find a way to shorten lead times and reduce cost. After examining the results, we gave top priority to shortening the implementation phase. The solution we settled on came from an assessment of the dynamics of project management. Like other companies, we had a lot on our plate, so every project had a tough schedule, with another development effort stacked immediately behind the current one. When development extended beyond management's tolerance, we declared victory, released the application for installation, and reassigned most of the team to build another system.

The application's premature birth gave the installation team fits. They nursed it along as best they could, but inadequate testing, insufficient user training, and process discrepancies made a job a long haul. Since the team had been cut, they didn't have time to fix every problem they encountered, so they queued them up as part of the maintenance backlog. We discovered our maintenance backlog didn't consist of newly requested changes, but comprised changes and modifications that rightfully should have been as part of original development. (Over half of our maintenance backlog was identified by users within the first 30 days of implementation.) Needless to say, these problems didn't endear us to our customers.

We rejected the obvious answer. Instead of extending development, we released even closer to schedule and instituted a Post Installation

Review. The Post Installation Review occurred 30 days after installing the system in test departments. The review included a thorough assessment of the application's fit with the organization, processes, and skill set of the users. We measured performance, scrubbed for bugs, and reviewed the adequacy of training. Resource scheduling provided the secret ingredient of our new methodology. We scheduled the development team back onto the project 60 days after release. They took a break from their new assignment and put a full-court press on fixing the things identified during the Post Installation Review. We turned a dromedary into a camel; instead of a single-humped schedule, we had a double-humped project cycle.

This new methodology worked better than our optimistic projections. User acceptance soared, projects were brought to closure, systems propagated with ease to the remainder of the organization, maintenance declined, department throughput increased, and benefits accrued to the business much faster.

How did we land on this solution? Not easily. My applications staff argued vehemently for postponing release until the application was ready for prime time. I feared the developers would never finish and I knew my management wouldn't buy millennium schedules, nor continue to accept late delivery. The real reason I rejected their arguments was another lesson I learned from Jack.

Jack always claimed that users didn't know what they wanted, but they sure knew what they didn't want. He taught me to get something—anything—in front of users as soon as possible. It's easier for them to say, "not like that, like this," than for them to envision an intangible system in the sky. I couldn't give schedule relief to my applications people, so the double development cycle provided the only answer. Throw it out there, see what's wrong, then bring it back and polish the rough spots. The method works beautifully.

Convincing business management that this was a sound approach proved even more difficult than selling the idea to my own staff. The strategy appeared to elongate follow-on development, which was held up as staff circled back to the prior project. They finally gave approval to test the concept once they understood that it really represented a timing difference in the application of resources, rather than an increase in cost and delivery schedule. Once experience demonstrated the advantages of a dual cycle development process, business leaders strongly questioned our judgment when we proposed a traditional schedule for small projects or enhancements. If you're not using a similar approach, try it; you'll be pleasantly surprised by the overall improvement in delivering applications to your organization.

Development Tools

Modern development tools reduce cost by increasing the productivity of your development staff. However, more money has probably been wasted on development tools than any other category of software. While some good tools are available, much of this software adds bureaucracy, commits you to a proprietary solution, or generates applications with slow performance.

Everyone searches for a repeatable process for applications development. This objective may be paramount in keeping your career on track, but you shouldn't adopt a development methodology that imposes a stranglehold bureaucracy, nor should you try the one approach I guarantee works—freezing technology.

Rapidly advancing technology and business processes changes dampen the learning curve required to drive a predictable process and continuous improvement. Backing off the overuse of leading edge technology helps, but you can't ignore new application architectures, tools, and methodologies and continue to be competitive. You want to adopt a standard methodology, but keep it simple and flexible. (The Software Engineering Institute's Maturity Model for Software specifies the general principles of a good development process.)

Take care in selecting your development tools. Make sure they generate fast applications, provide flexibility, achieve broad industry acceptance, and don't create legacy problems. Make especially sure your tools don't trade development productivity for reduced end user productivity. Development is a one-time event and users must live with a poor performing application for years. The right tools do increase productivity for application specialists, but tools alone never substitute for experienced, skilled professionals.

Once again, there are no simple answers. All management is dual thrust: You do the best you can within today's constraints, while broadening those constraints for tomorrow.

Team Size

I once had a development team working feverishly on a high-profile project. Despite the project being behind schedule, I had to reduce the team size from 16 to 12 because of economic conditions. To my surprise, the pace of development quickened. I learned a new lesson—the smaller the team size, the easier it is to coordinate among the members, which increases each individual member's productivity. Sometimes large projects require large teams, but a properly architected system can break the task into modules for small subteams. My experience shows that when teams exceed six people, productivity suffers from the need to coordinate and communicate between the members. Anything you can do to reduce team size takes complexity out of managing development, which reduces cost and schedule.

Speeds and Feeds

The technology industry desperately needs better tools to test for bugs, scalability, intuitiveness for mortals, and consistency with the process. Stress testing and destructive testing present such a problem, they are often omitted. Despite the lack of a comprehensive tool, point solutions exist and represent money well spent.

Due to added complexity, network computing needs performance modeling. In the absence of a good stress testing capability, performance modeling becomes crucial. Too many network computing applications die on deployment because they cannot scale to real-world volumes. Unfortunately, most performance modeling tools belong to the vendors who sold you the hardware or software. This conflict of interest doesn't present a major problem if you insist on contractual guarantees of modeled performance.

Cheaper, Faster

Conventional wisdom says that cost reduction comes as an automatic by-product of squeezing down cycle time. Marketing brochures for application tools claim their products not only shorten development time, but reduce cost. Cost and cycle time are closely coupled when you look at the total picture, including the business savings the application delivers, but it's not necessarily true when you look solely at applications development.

Different tools and methodologies have their preponderance of impact on either cost or schedule. We came to this realization as a result of the study I mentioned earlier. Time dominated the front and back ends of a project, while cost mounted during development or procurement. We decided we needed two sets of tools and methodologies to attack the issues simultaneously: one set to increase development productivity, and another set to shorten the design and implementation phases. A few quick examples will demonstrate my point.

Saving Time

Rapid applications development (RAD) provides a clean methodology for reducing cycle time. The best part is getting everyone in the same room and not letting them leave until they specify their requirements. First, this eliminates the meeting carousel, and second, conflicting requirements get on the table in front of everyone. Use RAD to avoid cycling back to meet with people because their compatriots down the hall want something different.

Prototyping allows users to see and experiment with an application, leading to a much better design. Although prototyping is used on the front end, its real benefit is saving time on the back end. A better design encompasses more of the business requirements and avoids a lot of scurrying around at the last minute to fix omissions. Make doubly sure your users know the difference between a prototype and a production system. Unless you manage their expectations, they'll be disappointed when it doesn't appear on their desk Monday morning.

A well-thought-out training plan can dramatically shorten the implementation phase. Training represents the biggest hidden cost of computer applications. Done right, training is expensive, done wrong, the cost becomes exorbitant. Inexperienced designers rely too much on the intuitiveness of a graphical user interface. Every good application includes process change, and process change requires training and indoctrination. Invest in training professionals and give them the tools and user attention necessary to do the job right.

Saving Money

Logically, purchasing applications should reduce your costs, but it doesn't shorten the front end or back end where most of the time gets spent. In my experience, shopping around for a ready-made with a decent fit consumes as much time as internal design. The perception that a project doesn't start until management approves the purchase creates an impression that buying is faster. Packaged software can elongate the back end because training and integration issues expand to accommodate conforming to the software's predetermined process flows.

Plagiarism provides the best technique for reducing cost. The car companies discovered that interchangeable parts reduced cost, and the same is true for software. You can accomplish this with Object technology. Objects are small pieces of code assembled into an application instead of writing entirely new programs for each application. Objects represent a logical extension of old-fashioned standard subroutines. Whether you aggressively push Objects or use simpler techniques to borrow prior work, reusable code saves enormous money and improves quality by reducing program bugs.

One caveat—reusable code requires astute and persistent management. Most developers readily use available code, but a few mavericks resist because they always think they can do it better. You need strict discipline to keep these people on program.

Not being able to quickly find what you want thwarts commonality. If it is easier to design from scratch than to use existing code, programmers will always write new code. To make reuse a standard practice, you need a code repository with good module descriptions and a flexible search engine. It's useless to have an extensive Object library with an inadequate catalog.

Contracted services present a good example of something that saves money, but elongates cycle time. Contractors always need to climb a learning curve for the organization and the project, but they provide a solution for the development bubble without carrying excessive payroll during the troughs. (A Gartner study found it took, on average, four months longer using outside services than when companies handled projects internally.)

Acquiring specialized expertise frequently requires using contracted services. We did a skills survey once, and discovered that all the new knowledge resided with our contractors. Don't get caught in this trap. Make sure you transfer the knowledge from your contractors to your own staff, otherwise your outside experts will eventually hold your applications for ransom.

Data versus Process

A battle has been waged incessantly between those who advocate a data approach to design versus those who insist that logic must drive design. Since every application includes both data and process logic, this argument will never be resolved. It's a chicken-and-egg debate. My design philosophy follows these steps:

1. Define the desired process.
2. Design the output from the system.
3. Determine if the output data already exists:
   - If yes, then get it from that source.
   - If no, then put it in an existing database, where it most logically belongs.
4. Design the input side of the application.
5. Define the internal logic.

Another Law dictates my approach:

Law #12   If data resides in two places, it will be inconsistent.

This series of steps makes life difficult for application designers. It forces them to search hither and yon for data, or requires them to unload, redesign, and reload existing databases. But consider the alternatives. You either end up with endlessly replicated data as each project takes a silo approach, or you undertake a gargantuan exercise to design an enterprise data repository.

Every time a major advance occurs in database management systems, the data proponents resurface intent on designing the ultimate data model. After some expensive failures, the logic proponents reascend to prominence and we await the next cycle.

Ideally, once in a while, separate from an applications design project, the data technicians should be given clearance to redesign a troublesome database from scratch. This effort should not include the entire free world, but be restricted to a specific database and ancillary files. Unfortunately, budgets, backlogs, and the lack of executive understanding make this a hard sell.

Software Redoubt

I think the business cry for "cheaper, faster, better" has caused some people to focus on only the first two elements. When this happens, it drives your computer initiative in a direction that inevitably leads to disappointment. The emphasis on the first two elements of the troika results in computer systems that are superficial, accident prone, and narrow-minded. In the network computing arena, designers sometimes discard the front and back ends of the development cycle to decrease delivery time. Inadequate design and lack of implementation discipline cause many big-time failures.

This situation evolved from management's need to accelerate process change and the network computing designers' limited experience. The small size of the initial network computing applications allowed a casual approach to design and the close proximity of technicians meant users could interactively work out design problems. Initially, users were tolerant because the technology was new, exciting, and a wonderful improvement over their previous systems. However, when network computing applications grow to enterprise scale, structure must be returned to the development process. Ad hoc design without proper emphasis on testing and implementation will not lead you to "better." It leads to disaster.

Applications still require arduous effort by the entire organization. If you agree the real business objective is to deliver applications that dramatically improve an organization's intrinsic might in the marketplace, then we need to reexamine the applications process and manage it better.

Whether you make, buy, or assemble your applications, certain principles remain constant throughout time and across technologies. Beyond the normal pitfalls encountered with anything new, system designers who lose sight of the fundamentals cause big problems. Businesses cannot afford to continue lackluster performance in applications, but the dismal performance of the past does not need to foreshadow the future. Managing your computer people effectively, and getting back to a few of the basics, can propel you ahead of your competition, especially if your competitors remain complacent.

Bill Gates said, "The revolution is here, and it is soft." All that hardware merely provides a vehicle for applications. Conceiving, designing, building, and implementing applications will always require astute management as we continue to test the limits of digitizing our processes and organizations. Corporate technology initiatives represent only one part of the triumvirate necessary to build a digital organization. You also need to select your technology with care and effectively manage your computer professionals.

JUNGLE WARFARE

Manage Your Supplier Relationships

Wrong principles badly applied will lead to frustration.

—HAROLD KOONTZ

Building a digital organization means managing a virtual organization. Interdependencies grow between organizations because no one can do it alone. Corporate infrastructure standards and the increased use of purchased software ties organizations to their suppliers. Service providers become increasingly important as organizations rent capabilities that used to be under internal control. Large segments of computer staffs comprise contractors or temporary help. Improved communications and lower cost have resulted in programming moving offshore. The rapid pace of technology creates a reliance on consultants and industry-watch firms. Whole segments of computing get outsourced to external providers. Increasingly, managing technology means managing a large number of supplier relationships. This requires a broader range of managerial skills than when organizations owned the entire function internally.

Selecting outside providers, setting up the relationship, managing the ongoing performance, and coordinating between internal and external services challenges the most astute executives. Since these relationships tend to be long term, you need to pick carefully, assure that agreements don't tilt in the suppliers' favor, never put these relationships on auto-pilot, and continuously monitor the interfaces and points of overlap between your retained and procured services. If you can juggle all this, you more than earn your keep.

Are these growing interdependencies good? I'm not sure, but they appear to be inevitable. Sometimes a new business structure drives interdependencies, other times reducing cost seems to be the driver, or perhaps the need for assistance in handling new complexities requires external relationships. You need to make your own decisions on what fits best with your organization, but no one can totally avoid establishing alliances, buying products that sway enterprise performance, or purchasing services critical to business continuance.

The Customer Is Always Right

You must negotiate every strategic relationship with self-interest as your supreme principle. Altruism does not exist on the other side of the table. Despite vows to the contrary, your potential "partner" really doesn't pine for a "win/win" relationship. This beguiling spiel serves to disarm your natural vigilance. After one salesperson too many said she wanted to establish a "long-term partnering relationship," I got a little ornery.

"That's fine," I said, "AlliedSignal has reservations about this technology and we need a partner to share the risk."

"You shouldn't be concerned. There's no risk with this technology. We have an excellent support team to help you overcome any problems you do encounter."

"That's not good enough," I said, "I want your quote cut in half so we can share the risk equally."

"I can't do that. I've already given you a steep discount from our normal corporate rates."

As I put on a bewildered pose, I said, "I don't understand, I thought you wanted to be our partner. A partnership means we both share the risk, the cost, and the rewards. If everything goes to plan, I'll pay you the other fifty percent."

Now it was her turn to look bewildered, "You're not serious, are you?"

"No, not really, but it's my money, so I'd rather be treated as your customer. AlliedSignal has a Customer Excellence program where we try to exceed the customer's expectations. That's the kind of relationship I want."

This smart salesperson quickly shifted gears and we started a good long-term customer/supplier relationship. You need to keep supplier relationships in perspective, remembering that profit motivates their behavior, just as it motivates your own behavior. Your suppliers want to increase profits and revenue. Just make sure that they accomplish this by providing superior products and services to your account.

Technology organizations use many different types of external providers of products and services. Exhibit 11.1 lists the broad categories of suppliers. You need a different approach for each category to maximize the benefit to your organization.

Technology Components

I've already described the Computing Technology Center's Gold, Lead, or Dead program. This program helps manage the ongoing relationship, but corporate standards cause a problem with initial procurement.

Setting standards, using common solutions, and consolidating purchasing power at the corporate level all leverage size to get better pricing and

---

Exhibit 11.1
Types of Business Technology Suppliers

Technology components.

Service providers.

Contractors.

Consultants.

Outsourcers.

Select Carefully, Negotiate in Your Self-Interest, and
Continuously Manage the Relationship and Interfaces

---

service. However, setting standards can be tricky if suppliers think they have a lock on your business. To avoid this hazard, make sure you really need a standard. Set standards only where consistency provides a measurable payback.

This first procurement step at the CTC includes a technical evaluation of the competing products. All the selected bidders make technical presentations back to back, with strictly enforced time limits. This information, along with appraisals from industry-watch firms and possibly lab test results, gets assessed by a technical evaluation team.

The technicians classify the technology into one of the categories shown in Exhibit 11.2. If no significant differentiation exists, then the product is declared a commodity, the CTC provides a list of the market share leaders to procurement, who compete the suppliers based solely on price. This situation provides the best opportunity for leveraging enterprise buying power, and despite the rare occurrence of commodities in technology, it still saves AlliedSignal several million dollars each year.

If some desirable differentiation exists, then the CTC puts a dollar value on the features in advance of requesting quotes. Procurement competitively bids the suppliers, adjusting the responses by the predetermined amounts. This works especially well for suppliers who bundle services or other premiums with the product. Salespeople try to convince you to adjust the quotes to the retail price of the freebies, but we always discount them to their near-term value to AlliedSignal. In fact, we frequently throw away the winner's bundled accessories if we believe they entice us into future commitments we're unwilling to make to the product.

Exhibit 11.2
Types of Technology Component Suppliers

Commodity.

Differentiated.

Unique or corporate standard.

The third category encompasses products possessing highly desirable or crucial attributes required for our standard platform. Whenever possible, delay publishing standards until after the bids. If only one product meets the specification, keep this knowledge to yourself. You must appear indifferent, otherwise suppliers will bid high, assuming the decision has already been made. AlliedSignal works hard to convince suppliers that we will change our standards if volume pricing doesn't meet our expectations. Sometimes we do a high-profile switch to send a message to the entire supplier base.

Convincing the salesperson that a huge follow-on order lies just beyond this procurement provides the best technique I've found to get good pricing. I've successfully used this technique over and over again, even with the same salesperson. Look at the company's entire product line and start a casual conversation about another product, dropping hints that you need a huge quantity of them. Another approach includes promising to do the next buy for another part of the world. This works especially well if the supplier needs a high profile account to penetrate other markets. You must sound credible, but I've always found some way to get salespeople so anxious for the next order that they give a better deal on the current bid.

Service Providers

Service providers include telecommunications carriers, equipment maintenance, leasing companies, disaster recovery, off-site storage, Internet access, value-added networks, and dozens of others required to run a modern technology organization. Service providers use all kinds of nefarious tricks to keep you from switching to another supplier. If you don't want your service provider taking you for granted, then review these types of contracts on a periodic basis to assure competitive pricing and quality of service.

The best pricing occurs with long-term contracts, but be careful if they bring you a smoking deal to extend an existing contract before expiration, or to switch from month-to-month to a multiyear commitment.

This type of marketing campaign almost always signals an eminent change in the industry. For example, local telephone companies give steep discounts for long-term contracts when independents start providing dial tone in the region. They want you locked in before you learn you have a choice. Likewise, leasing companies come calling to encourage early extensions when they discover a new technology destroys their residual estimates.

Ideally, you should never extend a lease. When the productive life extends beyond the term of the lease, purchasing provides a better alternative. However, productive life is a judgment call and we all make mistakes. You can never get good pricing on a lease extension unless you play the game right. Send a letter a few months in advance, asking the lessor to de-install and ship the equipment out. Close to expiration, inform the lessor that your boss wants a bid on extending the lease. Let them know this is merely a bureaucratic exercise and there is no chance of convincing anyone to keep this obsolete junk. They'll respond with a low-ball bid trying to convince your boss to override your decision to move to a later technology. Just in case the bluff doesn't work, be sure to secure a bid from another lessor for the same equipment. Then evaluate the lease extension against the disruption and cost of doing a "push/pull."

Incidentally, leases often include automatic extension clauses. Typically, they state that notification of cancellation must be given at least six months in advance or the lease extends automatically at the same rate. Don't waste negotiation points with this issue, when you sign the contract, hand over the cancellation letter at the same time. These clauses never say you can't do it immediately.

Creative Leasing

When we started building the enterprise network in Europe, we ran into tax problems that threatened our schedule. To meet the tax requirements of each country, we had to order the equipment in that country, or go through so many internal gyrations that the paperwork would destroy the Argonne forest. We wanted to deliver all the equipment to our French center, do the configuration and testing, then ship them out to the sites ready to plug and play. The company tax experts told us to do it country by country to avoid even the appearance of trying to circumvent national tax laws.

No way. We knew the site controllers wouldn't approve the requisitions without elaborate transfers of capital budget, local buyers wouldn't treat our requirements with priority, and we didn't have enough people to spray them all over Europe to do the installations. I told the network people to proceed with their plan and I would figure out a way around the tax issues.

After a few phone calls to other global companies, I received a reference to a Swiss firm that provided a service for this exact situation. They took ownership of the equipment, leased the boxes to the respective businesses, and acted as a true third party supplier. They had all the legal entities in place, were experts on European tax laws, and no elaborate, internal arm's length agreements were involved. Besides having competitive lease rates, the firm makes it easy to relocate equipment between countries and provides excellent asset management.

Ma Bell and Her Offspring

Quality of service should drive your long-distance carrier decision because one outage can cost your business more than the savings you gain by using a third-tier supplier. The telecommunications business competes aggressively, so you should be able to use your carrier of choice and still receive the best industry pricing. Nobody gives anything away, so you'll need to go through at least the early steps of a competitive bid process to get the best deal. Long-distance carriers use sign-up bonuses as an inducement for you to switch carriers. If you don't want to switch, but you still want the bonus, then you need to competitively bid the contract. Once you have a firm bonus offer, tell your current carrier you will stay with the company only if it matches the bonus.

The really big telecommunications savings for businesses in the late 1990s will come from local access. The breakup regulations for AT&T

never made sense. In the competitive long-distance business, Ma Bell continued to be tied down with Lilliputian regulations, while the Baby Bells used their local access profits to frolic in other people's neighborhoods. You can tell that pricing for local access carries big margins, because once the regulations were relaxed, everybody with a Yellow Pages listing issued a press release announcing that they were entering the business. Don't commit to anything long term for local access; prices will continuously free-fall for the next several years.

Luck Is No Substitute for Preparedness

Most companies neglect disaster recovery for network computing applications. The industry provides excellent cost, breadth of service, and methodologies for traditional mainframe computing, but has yet to establish similar capabilities for network computing. The risk grows as more mission-critical applications migrate to the new platforms. Your choices are limited. Organizations increasingly seek outside help for many computer functions, but my suggestion, at least for the time being, is for you to handle this one internally. You can't continue ignoring this business risk, nor wait for the pricing and service to meet your needs. An enterprise network, or intranet, provides a highway to connect replacement equipment in a different region, but the prerequisites include the off-site storage of data, quick access to servers and communications gear, and rehearsals of recovery procedures. You need to get on with fulfilling your responsibilities in this area.

Start with a business impact analysis (BIA) of your larger network computing applications. Then perform an audit to verify off-site storage of backup files. If a problem exists, then centralize the servers or impose strict guidelines. After these two relatively simple steps, you've already insured an ability to recover. With the requisite files stored away from your misfortune, a group of technicians can, sooner or later, get you operating again. Without the data, all is lost. The rest of the disaster recovery steps deal with the speed of recovery, as opposed to the ability to recover. These first steps should be your bare minimum, and I'll bet your audit committee doesn't know they haven't been taken.

Contractors

Contractors lay in wait for technology changes that become popular. I've seen this scenario played out over and over again: A technology breakthrough occurs or a purchased software package becomes popular; the early adopters train a cadre of technicians that learned the ins-and-outs the hard way; everyone wants the new technology, but no industry talent exists to implement it; and the few who have experience discover they can double or triple their earnings by jumping ship and becoming contractors. Organizations find they have no alternative but to pay exorbitant hourly rates to get the skills necessary to implement the technology. The cycle runs its course until supply meets demand, but another technology always overlaps, so the firms in this business never worry about future bookings.

The Big Six accounting firms, computer technology companies, and various independents retain people with differing skill levels to augment client organizations' internal staff. Although these firms call this service "consulting," it's really contracting, except when they provide guidance on architectural issues. They all bob and weave with the market, trying to keep their talent base skilled on the technologies in demand, but customers need to verify that all contractors have bona fide credentials and experience.

Make sure you transfer skills from your contractors to your own people. To preclude being stripped of your newly gained expertise, you might consider assigning some older analysts with heavy pensions in the offing. This goes counter to the habit of assigning younger computer specialists to new technology, but my experience says a mixed team with balance between unbridled enthusiasm and hard-earned experience produces the best results. Besides, lackadaisical performance often results from boredom, so give the seasoned professional a shot of adrenaline with a new challenge. They'll stick around and may very well surprise you.

Offshore programming increases in popularity. AlliedSignal owns an applications group in Ireland and uses firms in India for offshore applications support. While the cost in India is less, the Irish—as AlliedSignal employees—know the company's internal processes, work regularly with other applications groups, and can easily travel to European operations for on-site coordination and support.

The major requirements for successfully using offshore support include well-documented specifications, programming standards, consistent interfaces, and effective communication. AlliedSignal uses the Irish operation for more complex projects and usually has the Irish team leader participate in person during the design phase. This allows the team members to get to know each other and greatly improves later communication by e-mail, voice, and videoconferencing.

My best advice for using an outside firm for offshore support is to pick a good partner and stick with the company. There is a steep learning curve on both sides. Don't judge the effectiveness on initial results, wait until both sides learn how to coordinate across the globe. AlliedSignal's experience shows that with some experience, coordination improves and methodologies mesh between geographically dispersed teams. Retain external design responsibility, but allow your offshore team some flexibility—within standards—for internal design. If you specify everything, you might as well do it yourself because programmers write code faster than specifications.

Consultants

Good computer consultants have highly developed sensory perception and deftly slide their message along with the shifting trends and customer bias. I once asked a consultant I respected what skill was crucial to his field.

He answered unhesitatingly, "The ability to listen."

"Really, to whom?"

"The client, especially the prospective client. On your first visit, the client explains his problem, his constraints, and usually give hints as to his preferred course of action. You go back, put it all in a brief, and on the next visit you present it as fresh insight. The client thinks you're a genius because you think exactly as he does."

"Wait a minute," I said, "I thought consultants brought unique expertise to problems, or possibly opened management's eyes to alligators slinking into their pond."

"That's right, but first you have to get the engagement, and that's determined more by emotion than reason. If you want the opportunity to set things straight, you first need to convince the person with decision-making authority that you're the right consultant for the engagement."

These comments, from an extremely successful consultant, were an indictment not of the consulting profession, but of executives who shop for a second opinion to lay off some of the accountability for their decisions. Consultants can add real value, but the prerequisites include selecting the right consultant for the job, getting what you pay for, managing the engagement, and accepting valid recommendations and advice, even when it hurts (Exhibit 11.3).

Jack of All Trades . . .

Consultants develop deep expertise and become highly knowledgeable about their chosen specialty. Even general management consultants concentrate on organization design, business trends, globalization, or another business discipline. As a result, successful engagements in the past should not necessarily dictate your choice of consultants for the future. Examine the subject where you require help, and then search for the experts in that specific area. The high cost of consultants, along with their potential influence on the enterprise, warrants bringing in people with undisputed expertise in the subject of the engagement.

Exhibit 11.3
Managing Your Consultants

Select the right consultant.

Avoid "bait and switch."

Manage the engagement.

Receive recommendations with open mind.

When you whittle down the list, include a few professionals in private practice, especially if you're looking for a highly specialized expertise. Individual consultants tend to fit this description, but more important, what you see is what you get. Large firms often use industry-renowned experts to pitch the engagement, and then assign less experienced staff to do the work. You must avoid bait-and-switch tactics before you sign the contract.

When we looked for a consultant to help align computer technology with our business strategy, we had a clear first choice. However, we knew we would never see the senior partner again after we engaged the firm. We called him in and told him the firm had won the contract, with one condition. He smiled brightly, expressing his gratitude and told us we would not be disappointed. Then he asked about the one condition. His smile evaporated when we said he had to lead the engagement full time.

We both knew he had a huge backlog of other companies to reel in and no time to actually direct a specific engagement. We eventually compromised. He committed to spending a minimum of two days a week directing the engagement. It wasn't everything we wanted, but we knew that if he guided the rest of the team, we would get our money's worth. Everyone was pleased with the results, even the senior partner who thanked us for giving him an excuse to get back into the fray.

Define the Engagement

Provide clear directions for the engagement, secure contractual commitment on mutually agreed-on deliverables, and give parameters within which any solutions must fit. A fine line exists between spoon-feeding your desired answer to a consultant versus defining the constraints of your culture, business strategy, and budget.

Most consultants work very hard for their clients, but a small number make self-promotion their sole focus. The computer field includes a few who invent proprietary jargon to camouflage limited innovations in their offering. It is bad news when computer executives go to an industry meeting and discover they don't understand a word being said. Executives who make their living directing the introduction of new technology rightfully fear looking out of touch. There's no choice but for them to attend a high-priced seminar so they can understand the new buzzwords. The vast majority of consultants make important contributions, but if self-promotion and hype seem to be their primary objectives, then seek out someone who has their clients' interests at heart.

Evaluate Recommendations with an Open Mind

In a former position, prior to joining AlliedSignal, a consultant once came to me for advice. He had concluded that the right course of action would take the company in a direction different from the one many high-powered people wanted. I recommended he stick with his convictions. When he made his presentation, he encountered stony silence. Eventually, one of the executives asked if the report sitting beside him included his recommendations.

When the consultant answered yes, the executive said, "Hand it over and consider your engagement complete." The report never again saw the light of day, even though the consultant's recommendations proved correct, as we learned to our regret several months later.

I felt bad for my part in his public dismissal, but I learned a lesson. Some people don't want advice that runs counter to their preconceived notions. When you pay big engagement fees for an expert opinion, keep an open mind and honestly weigh the consultant's advice against your own judgment.

The Outsourcing Juggernaut

The popularity of outsourcing defies explanation. The seductive appeal persists despite the mixed reviews from existing contracts. The constantly improving structure of outsourcing arrangements accounts for part of the reason this industry continues its rapid growth. Technology executives keep working with the outsourcing model to plug the holes, while the industry adjusts its offerings to respond to changing market conditions and customer aspirations.

Outsourcing, as opposed to purchased services or contracting, occurs when the outsourcer takes ownership of something that used to be the customer's obligation—either assets or people—and then contracts to continue providing the service these people or assets used to perform internally. Outsourcing presents a tricky management challenge because you must negotiate the transfer, as well as the service and pricing. It also requires continuous vigilance during the contract. Allowing your outsourcer too much freedom is a serious mistake. Exhibit 11.4 lists the management tenets for outsourcing.

Put Yourself on the Other Side of the Table

Prior to starting down the outsourcing path, you need to understand the business from the outsourcer's perspective. First, they plan on making a profit, and they don't rely on being better than you. They target markets where they know technology will soon reduce the cost of provisioning the service. If they can get you locked in at prices assuming the current cost structure, then margins swell as they incorporate the new technology. That's why in the mid-1990s, they shifted from data centers to desktop support. Emerging technologies will soon reduce the cost of managing networks, workstations, and related infrastructure.

Besides betting on technology trends, outsourcers intend to lock you into their services using three techniques: They'll try to strip you of your entire skill base, they anticipate lax management, and they will try

---

Exhibit 11.4
Managing Outsourcing

Understand the industry dynamics.

Retain professional help.

Primary emphasis on commodity services.

Take cost out first.

Emphasize relationship over contract.

Maintain a competitive environment.

Retain internal expertise.

Structure for continuous price pressure.

Effectively Manage Ongoing Relationship

---

to make the cost of switching prohibitive. Awareness of these techniques is your best protection against an unsatisfactory relationship.

If you let the outsourcer hire your entire skill base, then you have limited opportunity to direct their work and the ability to re-insource becomes expensive and a long-term prospect. AlliedSignal retains architectural control over its outsourced functions and keeps a small contingent of experts to direct the outsourcers' use of technology. Besides putting pressure on the outsourcer to keep AlliedSignal contemporary, this group signals that the company retains insourcing capability. Always be sure you own any licensed technology used on your account and get contractual rights to use any outsourcer developed technology in the case of contract termination.

Seek Professional Help

Since outsourcing keeps moving with the market and customer demand, I can't give you exhaustive advice in this area. You shouldn't attempt a major outsourcing initiative without professional guidance. You will never be an expert at outsourcing because it is not one of your core processes. Setting up deals is a mandatory skill for outsourcers and they have special teams adept at negotiating agreements. You can learn outsourcing trends, pitfalls, and best practices from consultants and lawyers who exclusively practice in this narrow specialty. Outsourcing conferences also provide a good baseline, but be aware that many of the presenters represent outsourcing interests.

Don't Let Outsourcing Become Single Sourcing

Commodity services provide the best candidates for outsourcing. Anything that you can buy from any number of sources lowers the risk of selecting a poor performing supplier. Venturing into specialized business processes may lock you into a single supplier. Even if performance meets your expectations, the limited pricing pressure will increase your cost over time. If you have never outsourced previously, it is especially valuable to start with something relatively simple.

Outsourcing the routine runs counter to what many companies desire. They want to unload their biggest problem, in the false hope that the outsourcer has some special skill to resolve the difficulties. Rapid change causes the major difficulties in adopting computer technology. Outsourcers do not have a magic answer for the absence of a skilled labor base, nor do they typically own proprietary solutions to fill the holes in the technology.

Approach it this way: If you have a severe technology problem, outsource the routine so your internal brain trust can concentrate on developing a solution to that problem. If you're having this problem, probably everyone else is encountering the same difficulty. Problems common across industry spur the suppliers of technology to fill the void with new products. When you outsource prior to a generally available solution, your service provider reaps the benefits as advancing technology resolves the issue.

Fix It, Then Outsource

If you turn a garbage dump over to an outsourcer, the firm will rename it a landfill and push the problems around. If you make a strategic decision to outsource a mismanaged function, clean it up first or you'll continue to see poor results at expensive rates. Outsourcing doesn't succeed when it represents an "I give up" strategy. Put your best talent on the area and execute a fast get-well program. Rip cost out, smooth out your processes, update your equipment, and then bring the outsourcers in to bid.

One more point, move your best people out of range before contacting the outsourcers. Move in your talent, get the job done, then reassign them to your next biggest problem area. You didn't put them on this assignment to fill your outsourcer's recruitment needs.

Terms and Conditions

If you rely solely on contract language, you'll be disappointed. Outsourcing really does require a long-term partnering relationship. If the relationship gets defined purely on legal language—you lose. You want a "partner" intent on drawing you in with products and services, not one determined to entrap you with contract language. If your selected outsourcer seems obsessed with the contract, then dump the company and find one who wins market share with superior performance.

Compete the Edges

Maintain some level of competition so you aren't totally dependent on the outsourcer's goodwill. Providing competition presents difficulties because the outsourcer normally has the entire process under contract. Lots of theories have been proposed to maintain some semblance of competition, but the only one that I've seen work is competing the edges of the contract with either internal resources or other outside suppliers.

Managing the Contract

Retain enough expertise to provide second opinions and aggressively manage the outsourcer's performance. Establish a culture that says you cannot move into higher executive positions until you do a stint managing an outsourcer relationship, then rotate your high-potential people through the position. Establish comprehensive metrics and distribute them on a regular basis to general management. Never let the outsourcer believe the firm's performance doesn't have the rapt attention of your entire company's leadership team.

When the outsourcer assigns an account executive, don't merely exercise veto power. Require the outsourcer to present three candidates for you to choose from. With a yea or nay vote on a single candidate, the tendency is to accept the executive unless some glaring fault surfaces during the interview. By selecting one of three, you're not making career limiting value judgments on the other candidates.

Keeping Pricing Competitive

There are many ways to maintain competitive prices during the term of the contract. Keeping the contract term as short as possible provides the best protection. Interim techniques include benchmarking costs against new contracts, "most favored nation" clauses, scheduled rate reductions during the contract term, tying pricing to industry indexes, and several others. When you engage your outsourcing consultant, he or she will direct you on the best techniques.

Core Competencies

Outsourcers market their services on the basis that if the process isn't a core competency, then it distracts management's attention from the important business issues. The concept is that you're better off letting someone else handle it that does have the process as a core competency. I think this frames the question improperly. I believe you should ask yourself what the business impact would be if the process was performed less than adequately. If you cannot endure poor performance and stay in business, then you should retain the process internally. The risks are too high.

Many technology functions fall into this category. Your processes dictate the way you do business, and your computers dictate your processes. I believe the wholesale outsourcing of the entire computer function will eventually inhibit an organization's competitiveness. Your destiny rests in someone else's hands. My advice is to use care when outsourcing: First select the most routine for consideration, negotiate a balanced agreement, actively manage the relationship, and retain some expertise in the outsourced function. You can't escape numerous interdependencies with other organizations, but use business sense when you make these decisions. Retain the crucial as you off-load the mundane, the routine, and the commonplace.

Manage Your Supplier Relationships

Whether you decide to outsource a major portion of your technology function, or selectively use outside service providers, its certain that managing technology requires establishing numerous supplier relationships. Technology suppliers will continue to provide crucial services to every organization. I started this book by saying you need a good strategy. As you fashion your plan, decide early which functions need outside assistance and select your suppliers with as much care as you use to staff your immediate organization. A digital organization need not be elusive, if you partner with the best the computer industry has to offer and manage the relationship.

EPILOGUE

The Continuing Saga

You ain't heard nothin' yet, folks.

—AL JOLSON

What does the future hold for technology? Omnipresent computers. Big ones, small ones, even tiny ones you never see. Ones that talk, ones that listen, even some that seem to think. They'll be everywhere and function so effortlessly, you'll believe these machines read your mind. Everyone and everything will be connected—all the time. The most important innovation since the assembly line will permeate and intrude on every aspect of our lives. The premier vacation of the future will be an unplugged oasis, where no one and no machine can touch you.

The competitive marketplace fuels the information revolution. The captains of our most important industry carefully craft offensive strategies and elaborate defensive plans. Computers capture the hearts and minds of our best and brightest before they're even old enough to drive. The kind of people who were composers and writers in previous centuries now write software to entertain, educate, and ease our work life. Where the industrial revolution provided humankind with tedious jobs and mass quantities, the information revolution captivates its disciples and brings back tailor-made products.

The computer amplifies its influence beyond the environs of its own industry. Every scientific discipline explodes with discoveries by using this invigorating new tool. Knowledge expands exponentially as scientists, academics, and thinkers of all kinds leverage their brains with these fascinating machines. Everything that is now known will look shallow in the near future. Are there downsides to all this technology? Of course, but as always, people will muddle through and improve human existence. Change is difficult to accept, but it has become a permanent fixture of the modern world.

The only thing that will not change is the need to effectively manage the adoption of new technology. The computer industry does an excellent job of bringing brilliant new products to market, but we, as managers, need to do a better job of integrating the technology into our organizations. Some organizations will improve their skill in adopting technology faster than others, and they will gain the elusive competitive advantage. World competition is fierce and only digital organizations will have the wherewithal to survive in a global economy.

Navigating the Future

The theme of this book has been to apply good management fundamentals to computer technology. My advice includes understanding the industry, surrounding new technology with a workable infrastructure, choosing your technology with care, and managing your computer people. Your organization's investment in technology will continue to grow as a percentage of sales, so you must manage it as effectively as the rest of your business. The rapidly changing future holds promise, as well as uncertainty. Navigating around the obstacles will never get any easier, so assign your very best executives and managers to your technology function.

As computer technology continues its explosive growth, the industry will reshape itself, new disciplines will emerge, and technology fads will surge and wane. I'm not a futurist, just a simple operations person who makes technology work. At the risk of getting out of my league, let me try my hand at predicting some of the future macrotrends for the technology industry.

Purchased Software for Business Processes Will Fade

Let's face it, if everybody uses the same software and processes, where is the competitive advantage? The ideal solution would be the quick assembly of custom software from standard components. The software discipline is proceeding along this path. Shrink-wrapped software will always exist, but the programs that dictate business processes for large organizations will be custom tailored within your four walls. Don't stop buying process applications; the half-life of business software is only five years and it will take a few more than that to make standardized components a reality. And don't fret about the business application software houses—they'll be the suppliers of your building blocks.

Telecommuting Will Grow and Then Contract

Many people initially will be attracted to telecommuting, but then will pull back into a central work environment after they discover human communication uses all five, and perhaps six, senses. Some will continue to work at home but they will be the information age equivalent of a cottage industry. Home-based workers will represent two extremes of the business continuum: either doing mundane tasks, or highly creative, individual projects. The rest will increase their productivity by interacting daily with the rest of their team.

The Internet Will Become Omnipresent, Invisible, and More Expensive

The next great growth area for the Internet will be devices. Cars, home appliances, business equipment, and personal devices all will be interconnected through the World-Wide Web. As everything and everyone becomes connected, the mechanics of the Internet will recede into the background. The financial subsidies for the Internet, both government and private, will gradually be withdrawn, forcing users to pay a greater share of the cost. The Internet will deliver television programming, but it will not supplant the content. The Internet will invent new multimedia, interactive content that exploits the characteristics of the media. As through the ages, storytelling will remain the province of the storyteller, not the listener.

Outsourcing Will Evolve into a Utility Business

Outsourcing will evolve closer to the service bureau model as organizations discover their own people do a better job at the "up close and personal" aspects of computer technology. Data center economies of scale will continue to shift dramatically to the right with new resource management tools and ever larger machines. Outsourcers will operate massive centers providing utilitylike services for organizations that will increasingly off-load their central computing to these service providers.

The Network Computer Will Fill a Niche Market, but Not Replace the Personal Computer

Some businesses will adopt the network computer to simplify their technical environment, but once achieved, there will be a gradual migration back to the more flexible personal computer. The integration of browsers into the operating system, automatic Web interfaces for applications, and a small price differential will accelerate the demise of the network computer in business. In the home, consumers will balk at paying each time they play a game or use a shrink-wrapped application. The network computer will win limited home market share as a second machine dedicated to surfing the Internet.

Value-Added Networks (VANs) Will Flourish

Everyone predicts the death of value-added networks and proprietary online services. Smart executives manage these service providers and they will think up new ways to continually add value to the raw Internet. Expect to see some interesting and high priced acquisitions of Internet providers by the VANs and proprietary online services.

EPILOGUE

Everyone Will Use Multiple Personal Computers

Many personal computer users have machines at work, at home, and one for the road. Separate, multiple machines will continue as the norm, expanding to include more of the population of computer users. However, the computer industry will soon make them as seamless to switch between as walking between rooms tuned to the same TV channel.

Mainframes Will Survive, as NT™ Explodes, and UNIX Regresses

After other operating systems copied UNIX's vital parts, it has little more to offer the business community. A reawakened IBM will recognize MVS as its crown jewel and continue to develop it as a enterprise-scale server. Microsoft's NT will capture everything else but technical computing, where UNIX will continue to reign.

Globalization and Competition Will Cause Crib Death for Most Baby Bells

The Baby Bells and AT&T's domestic long-distance competitors will merge, form equity alliances, divest, and acquire in a furtive attempt to stay in the game. Local access competition will suck the vitality out of the Baby Bells, while competition from European telecommunications companies will weaken second- and third-tier long-distance carriers. Globalization will cause an industry contraction to a few world players and many also-rans. The victors will include AT&T, British Telecom, and consortium from the European continent, and a new player made up of previously independent U.S. carriers. The two spinoffs from AT&T will slowly lose market share and industry presence.

Home Shopping Will Always Fill Only a Niche Market

Video movie rentals didn't hurt movie theaters and shopping from home will not turn shoppers into stay-at-home recluses. In both cases, the experience is different, and leaving the domicile to mingle fulfills a human need for affiliation. Home shopping for government services will grow because grappling with government workers isn't always a pleasant experience.

IBM Will Resurrect Itself as a Benign Industry Leader

The sleeping giant has awakened. IBM now differentiates its products, as it concentrates on its strengths and challenges competitors in its core business. Bill Gates, going on to other business, believes he slew the dragon, but IBM is sneaking up from behind. Big Blue will lose some battles in the marketplace, but it will capture a dominating share in services, system integration, and enterprise-scale technical solutions.

Application Builders Will Become Business Generalists

Real programmers will work for software houses developing shrink-wrapped applications, development tools, and software components. Organizations will concentrate on hiring people with process knowledge. These new-breed systems analysts will possess business sense and good interpersonal skills, not deep technical knowledge. The deeply technical will cloister in specialized enclaves developing increasingly complex programs for the rest of us.

I'll be happy if half these predictions come true. Your job is to figure out which half. Whatever happens, smart organizations will leverage the power of computers to bring eye-popping products and services to market, change as they adapt to new technology, and continuously improve their skill at using computers.

The Digital Organization

AlliedSignal's goal is to "become a premier company, successful in everything we do." Larry Bossidy defines this as a quest. We have not yet achieved "premier" status in Bossidy's eyes, nor are we today the ideal digital organization. We did start toward the end of the pack, sprinted to catch up, and are now passing the competition as the training, equipment, and processes take hold. No one stands still, so staying competitive requires perseverance, fortitude, and rock-solid management.

I once asked one of my better project managers how she had pulled off a semimiracle. She told me, "the usual way."

EPILOGUE

"Baloney," I said, "I know you. Don't try and make it sound easy. You always take your commitments seriously, pick good people, lead them out of tough situations, make smart decisions, work hard, and you never give up."

"That's what I said, the usual way."

Too many people try the unusual. Making technology work for you will always be hard work. I said in the beginning it would take a great pile of silver BBs to resolve our technology issues. My intention was to help you stockpile the ammunition you'll need to build your own digital organization. I trust I haven't slipped in too many duds or blanks. All I can tell you is that the tips, techniques, and approaches I've outlined in this book work for me. I sincerely hope they also work for you.

APPENDIX

The Natural Laws of Computing

1. Technology breakthroughs require a surrounding infrastructure.
2. Enterprise solutions must be managed on an enterprise basis.
3. Things break!
4. Change causes the most downtime.
5. Industry standards inhibit innovation.
6. Market share wins, not technical eloquence.
7. Competitive advantage is hard to gain and maintain.
8. The scope of every computer project grows.
9. New computer technologies unveil additional layers of applications that suddenly become feasible and cost-effective.
10. Size is the greatest determinant of implementation difficulty.
11. Serendipity does not apply to computer systems.
12. If data resides in two places, it will be inconsistent.

INDEX

INDEX